GCSE

Geography

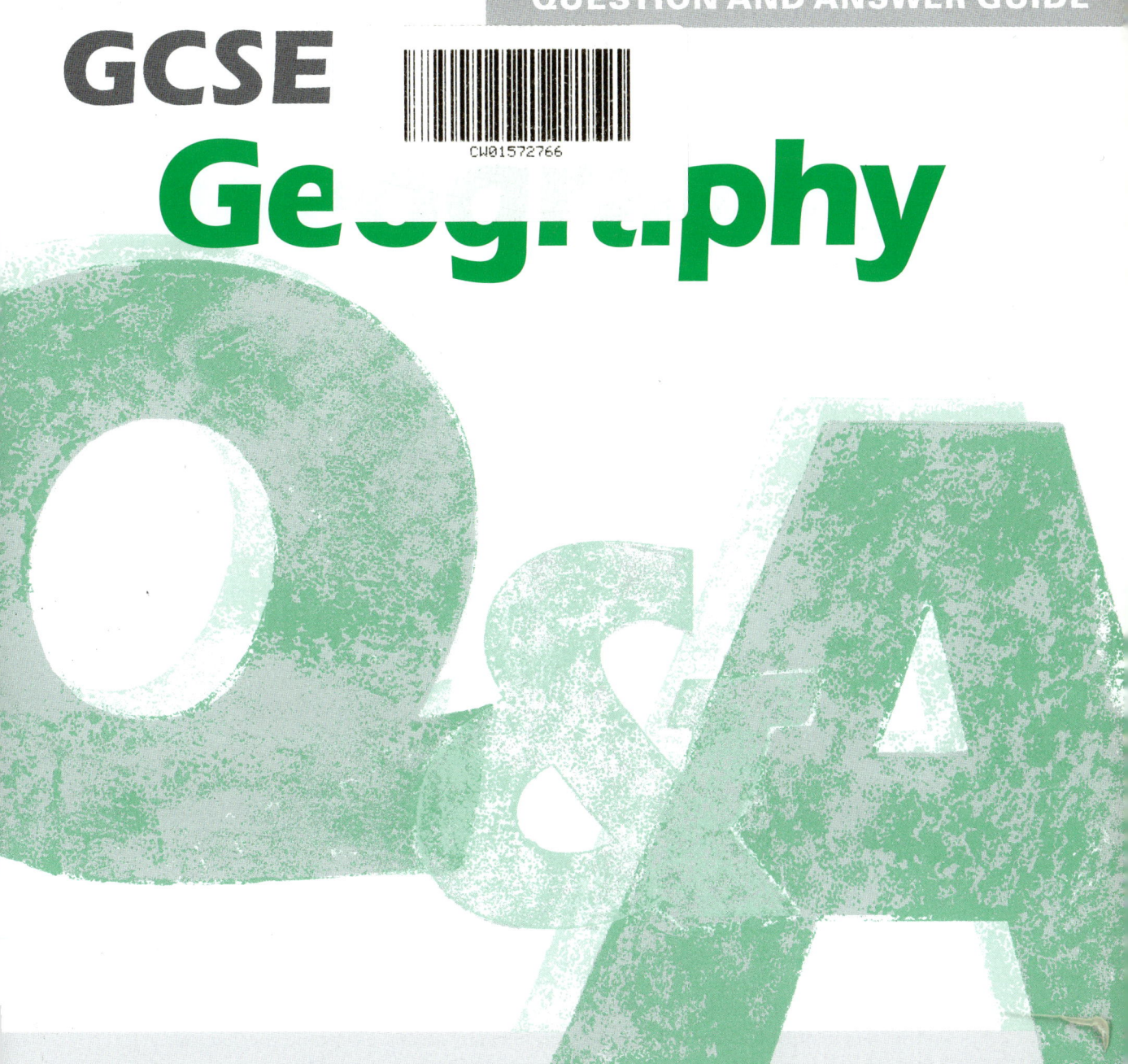

John Pallister

Philip Allan Updates
Market Place
Deddington
Oxfordshire
OX15 0SE

tel: 01869 338652
fax: 01869 337590
e-mail: sales@philipallan.co.uk
www.philipallan.co.uk

Cover illustration by Neil Fozzard

Printed by Raithby, Lawrence & Co Ltd, Leicester

Contents

Introduction

About this guide

The aim of this guide is to help you achieve a good grade in your GCSE geography written examinations. It concentrates on the most common topics included in the geography courses. Whatever GCSE course you are following, the majority of the topics covered by the 20 questions and answers will be included in it. Physical topics are covered first, followed by human and environmental.

The guide will help you to make the most of your knowledge and understanding, and thereby improve your grade. It does this by:

- giving examples of answers to different styles of GCSE geography examination questions
- pointing out what examiners are looking for when they mark these answers
- passing on advice to help you answer the question set in the most effective way
- highlighting some of the mistakes made by candidates in GCSE examinations in the hope that you can learn from these

A mixture of short and long questions is used for every topic, so that examples of the many different types of questions asked in GCSE geography examinations are covered. These questions are marked out of a total of 12, 15 or 20 marks. If the examination papers for your course have a different number of marks per question this does not matter, because GCSE grades relate to the *percentage* of marks obtained. As a general rule, you need 75% or just under of the total marks for an answer to reach grade A standard, and 50% or just over to be certain of achieving a grade C. Therefore, the comments made by the examiner in this guide about the content of an answer and its overall worth apply to all GCSE geography courses.

How to use this guide

Each section of this guide begins with a GCSE question of a type that may be used in any exam. You are advised to study it before reading the answers. If you have just covered the topic in class, or if you are preparing to take an examination soon, it is a good idea to write down your own answers to the different parts of the question before continuing.

Each question is followed by answers by two candidates, A and B:

- Candidate A's answers are generally A-grade standard, sometimes A*. The A* grade is a reward for consistent performance across the whole examination, which in GCSE geography typically consists of two written examination papers and coursework. In individual questions, you should be looking to achieve the equivalent of 17 or 18 marks out of 20.

- Candidate B's answers are generally C-grade standard. You can make mistakes and answer certain parts of a question badly, but provided your total mark reaches above half marks, you stand a high chance of gaining at least a grade C.

The answers are accompanied by examiner comments. These are preceded by the icon **e**. Marks are awarded for the answers to each part of the question and the marking is explained by the examiner. Comments by the examiner also highlight good points in the answers; any weaknesses are explained and improvements suggested. A summary of the worth of the overall answer is included at the end.

At the end of each question section the main messages from the examiner comments are summarised as exam tips. This will be helpful when you come to write your own answers under exam conditions.

If you are aiming for grade C, you are still advised to look at the answers of grade-A quality and take note of the examiner comments about them. They illustrate examples of good practice and technique. Many candidates who receive a grade C overall in the examination produce one answer of grade-A quality, without being able to maintain that standard throughout.

If you are aiming for grade A or A*, you are still advised to look at the answers closer to grade-C quality. Grade-C candidates make more mistakes and write a higher proportion of weak answers than those gaining higher grades. If you are aware of the common mistakes and study the examiner comments about these, you should be better able to avoid answers that reduce your overall level of performance. It is wrong to believe that candidates who gain grade A in GCSE geography never write a poor answer; it is quite usual for one or two significantly weaker answers to be compensated for by several particularly good answers. Similarly, every answer does not need to be perfect to gain a grade A*; they just need to be very good overall and better than those of most other candidates.

GCSE geography questions

Each examination question contains at least two parts (a *command word* and a *question theme*). Some questions also contain a third part (*where*). You must learn to recognise these. The better you understand the form of questions, the greater is your chance of giving effective answers.

Command words
This part of the question tells you what you need to do. The most commonly used command words in GCSE geography questions are given below:

Name.../State...
These are the simplest command words. Only short answers are expected. Often these are used in 1- and 2-mark questions.

Describe...

This is one of the most widely used command words. It tells you to write about the characteristics of a geographical feature, or to write down what can be seen on a graph, map, diagram or photograph. What it definitely does *not* command you to do is to explain. The length of answer expected is determined by the number of marks; you need to write more fully about what you can see on a photograph when there are 4 marks for the answer than when there are only 2.

Define.../What is meant by...?

These, and other similar command words, are asking you to give a definition of a geographical term or phrase. In examination questions you are most likely to be asked to define geographical terms that are stated in your GCSE course. Once again, the number of marks attached to the question determines the length of answer needed.

Explain.../Give reasons for.../Why...?

You are expected to use your geographical knowledge and understanding to *account for* the appearance or location of physical and human features on the Earth's surface. In these questions there will not be any reward for just describing. This type of command word is used in the majority of longer GCSE examination answers (worth 4 marks or more).

Question theme

This tells you what the question is about. For example, in the question 'Explain the formation of an oxbow lake', the question theme is *the formation of an oxbow lake*. The command word is *explain*.

In the question 'Describe the characteristics of the CBD', the question theme is *characteristics of the CBD*. The command word is *describe*.

These are examples of two-part questions consisting of command word and theme. You can write about the formation of oxbow lakes and the characteristics of CBDs for anywhere in the world.

Where

A place or part of the world is named in the question and you need to refer to it in your answer. For example, in the question 'State the problems for farmers in LEDCs', *state* is the command word, *problems for farmers* is the theme and *in LEDCs* is where. When a place or part of the world is included in the question, it is a vital part. You could write a brilliant answer about the problems for farmers in the UK and receive no marks for it because the UK is not an LEDC.

All parts of the question are of equal importance. If you ignore or misread one part of a question, you are likely to drift into an answer that is either partially or totally irrelevant and you will lose marks.

The importance of reading the question carefully

Under examination pressure, some candidates think they are writing a correct answer that matches perfectly the question set when, in fact, they are writing an answer to a different question. Look at the example below.

Question: State the causes of global warming.

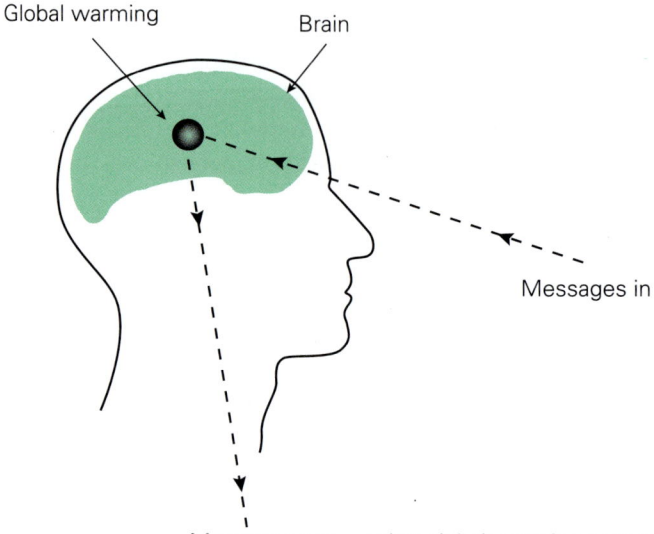

- The candidate scans the question.
- The message — *what global warming causes* — is transferred to the brain.
- The message goes to that part of the brain holding information on global warming.
- The brain instructs the candidate to write about what global warming causes.
- The candidate writes the following answer:

Global warming causes rises in world sea levels. Low-lying countries like Bangladesh will be flooded first. All low-lying coastal areas are at risk. In the UK the fertile farm land in the Fens will be flooded first.

The candidate is probably so convinced that he/she understands the question and knows the answer that he/she does not consider it necessary to look at the question again. Candidate's thoughts: *A job well done. One question in the bag. On course for a grade A.*

Disaster might have been prevented by:
- a more careful and closer reading of the question the first time
- a second look at the question
- a final check after writing the answer and before moving on to the next question

This might have led to the happier situation illustrated below.

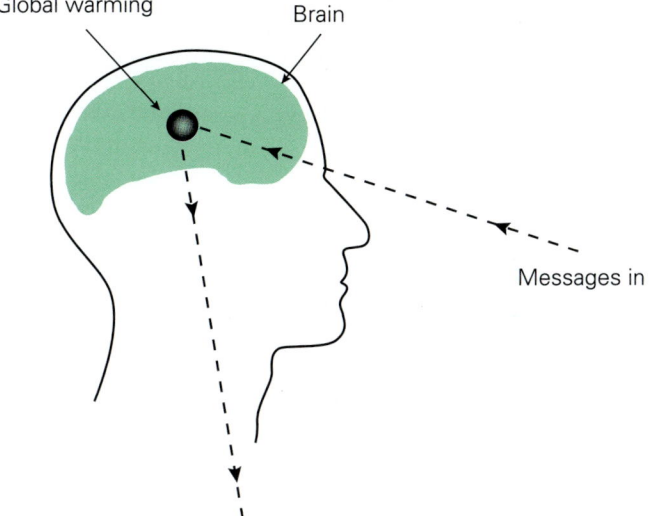

- The candidate carefully scans the question. He/she then reads it again.
- The message — *causes of global warming* — is transferred to brain.
- The message goes to that part of brain holding information on global warming.
- The brain instructs the candidate to write about the causes of global warming.
- The candidate writes the following answer:

Global warming is caused by the accumulation of greenhouse gases in the atmosphere. The main one of these is carbon dioxide from the burning of fossil fuels. Light from the sun passes through the atmosphere normally; then the greenhouse gases trap the Earth's heat, leading to global warming.

This is the correct answer to the question. Have you got the message? Read the question *very* carefully.

River landforms and processes

(a) Study the photograph below of High Force on the River Tees, the highest waterfall in England.

(i) Describe **three** other features of the river and its channel that can be seen in the photograph. (3 marks)

(ii) Name **two** ways in which waterfalls can be useful to people. (2 marks)

(iii) Explain how a waterfall is formed. (5 marks)

(b) (i) For **one** example of a river flood you have studied, describe the causes and effects of the flood. (6 marks)

(ii) Name and describe **two** methods for trying to prevent rivers from flooding. (4 marks)

Total: 20 marks

Answer to Question 1: candidate A

(a) (i)   1. The river bed above the waterfall is rocky and wide.

2. The river flows in a narrow channel where the waterfall is found.

3. The river water is fast flowing where it rushes down the waterfall into the plunge pool.

e These are clear, accurate answers, based upon describing what can be seen in the photograph. 3 marks.

(ii) For electricity (HEP) and tourist attraction.

e Two ways have been named. No extra detail about them is needed when the command word is 'name'. 2 marks.

(iii) Waterfalls are formed where hard and soft rocks outcrop in a river. The hard rock is more difficult for the river to erode, whereas the soft rock is easily eroded. The force of the water as it flows over the waterfall erodes the soft rock underneath it by hydraulic action and forms the plunge pool. The falling water splashes back and this helps to erode the soft rock behind the waterfall, so that the hard rock above is left overhanging. This later collapses which means that the waterfall retreats upstream. I have shown what happens on the diagram below.

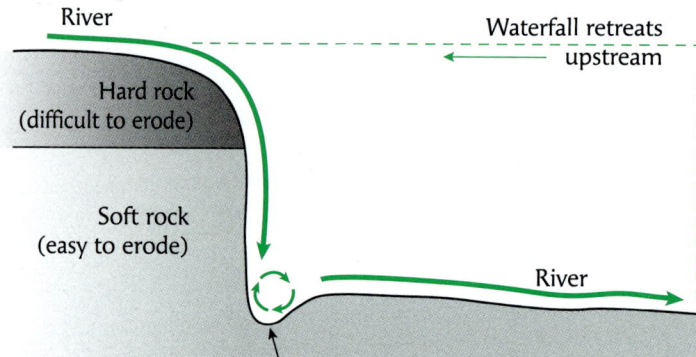

Plunge pool eroded by the falling water

e This is another good answer for the following three reasons:

- The candidate focuses upon explaining how a waterfall is formed throughout the answer.
- The answer is well arranged because it follows the order in which the formation of the waterfall occurs naturally.
- A reference to river processes is included, in this case hydraulic action, which is important when explaining the formation of landforms.

This answer would gain all 5 marks, even without the diagram. Examiners look at diagrams for extra points that are not covered in the written part. In this case, although the diagram confirms that the candidate understands how a waterfall is formed, all that was needed for full marks was already included in the written answer.

(b) (i) The river flood we have studied is the Mississippi. It flooded badly in 1993, killing almost 30 people and making 36,000 people homeless. It also flooded more than 2 million hectares of good farmland. This happened mainly in the north between the cities of Minneapolis and St Louis, where everything was under water for several weeks. The maize crop was ruined and animals on the farms were drowned. Everywhere was covered in mud by the flood and they estimated that it cost 10 billion dollars to clear it all up. In fact, the area affected was larger than Britain.

e This is a good answer about the *effects* of a river flood. The example is named first, which is always the best way to start answering a case study question. The effects are clearly related to the named example and specific information is given.

But there is nothing about the *causes* of the flood. Was this because the candidate was too careless in reading the question and missed the reference to causes? The answer would gain only 4 of the 6 marks. It would not reach the top level of marking because only one part of the question has been answered. The causes of the Mississippi flood in 1993 were a combination of heavy rainfall, lasting over a period of about 50 days, melting snow in spring and early summer thunderstorms.

(ii) The most effective method is to build a dam. The second method which is used on many rivers is to make the banks higher. Many rivers have high banks called levées. People can increase the size of these by adding more soil, stones and rocks to make embankments so that the river can flow at higher levels before spilling over the banks and flooding.

e Two methods are named — dam and high banks. However, only one is described — high banks. The way in which high banks can help prevent floods is well described, but the candidate should have described how a dam can, for example, hold back water during periods of high rainfall and prevent flooding downstream. It is likely that an examiner would award 1 mark for dam, and 2 marks for high banks, giving a total of 3 out of 4 marks.

Summary

What mark would the answer be given?
17 out of 20. This would be a grade A, probably an A*.

How could the answer have been improved?
By ensuring that answers covered all aspects of the question set — both **causes** and **effects** for flooding, both **naming** and **describing** for methods of flood prevention.

Answer to Question 1: candidate B

(a) (i) I can see a waterfall and the water is fast flowing. This is because the river is flowing over a hard outcrop of rock, which it cannot erode as fast as the soft rock below. The fast flowing water is forming a plunge pool that can be seen at the bottom of the waterfall.

    e Only the first sentence of this answer describes what can be seen in the photograph. The use of 'because' in the second sentence shows that the candidate is beginning to explain, which is not needed and for which there are no marks. Can the plunge pool described in the third sentence really be seen in the photograph? The candidate is using knowledge rather than observation and so has failed to answer the question set. 1 mark.

(ii) Big waterfalls attract lots of visitors such as the Niagara Falls. HEP is a renewable type of electricity which doesn't cause global warming.

    e Two ways are named so 2 marks would be gained. The answer gives additional good information about each one, but the question total was only 2 marks and it did not ask for examples, nor was further elaboration needed. However, the candidate was erring on the right side by giving the extra detail.

(iii) I have drawn a diagram to explain how a waterfall is formed. One example is the Niagara Falls. There is a long gorge below the main falls caused by the waterfall retreating backwards as the soft rock is eroded by the river. Types of river erosion are corrosion, corrasion and hydraulic action.

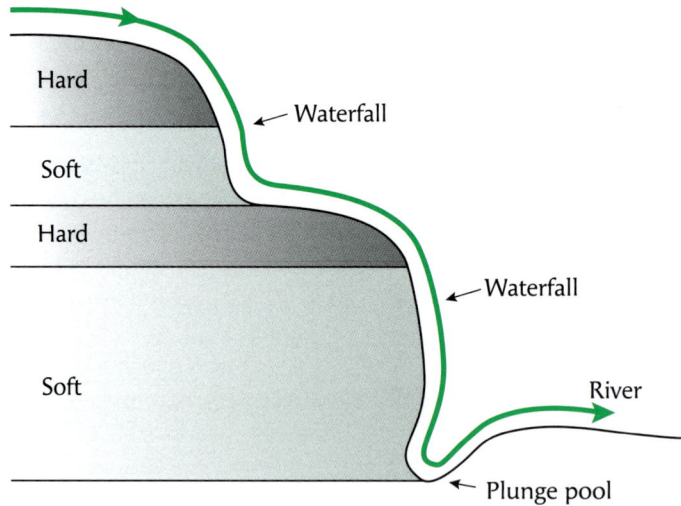

    e The written part of the answer is brief. The types of river erosion are just listed, instead of being used to explain the formation of the waterfall. Most of the labels on the diagram describe rather than explain, but some credit can be given for labelling the hard and soft bands of rock because this is not referred to in the written part. Overall, this is an incomplete answer which would be worth 3 out of the 5 marks.

(b) (i) The River Ganges floods often in Bangladesh. This is caused by heavy rain. The Ganges fills up with water from its tributaries and floods its banks. Many people in Bangladesh are farmers and their crops and animals are destroyed.

 One cause (heavy rain) and one effect (upon farmers) are stated, but there is nothing else. This is a low-level answer which would earn 2 marks; there is not enough content to earn more. At least the candidate did answer in relation to a named river.

(ii) The best method of stopping flooding is to build a dam. When it rains a lot, the water is stored in the dam and released during dry times of the year. This stops the land from flooding; an example is the Aswan High Dam on the River Nile. Another method is to keep the river bed free from sediment. This is done by dredging. When rainwater reaches the river, it can flow away faster.

 This is the candidate's best answer. Two methods are named and both are described, just as was required by the question. 4 marks.

Summary

What mark would the answer be given?
12 out of 20. This is a grade-C answer, comfortably above the C/D boundary. Good answers to some parts compensate for weak answers to others. A reasonable understanding and knowledge of rivers is shown.

How could the answer have been improved?
By the candidate using his/her knowledge better. This could have been done in two ways:
- Making a better match between the amount written and the number of marks. More than was needed in relation to the number of marks was given in (a)(ii), but not enough was written for the marks available in (b)(i).
- Paying more attention to command words. When asked to describe from a photograph, state only what can be seen and say nothing about how it was formed.

Exam tips

1 Obey command words.

2 Look at the number of marks — they indicate the length of answer expected.

3 Diagrams are always welcome in geography answers, even when not directly requested in the question, but to gain extra marks they need to give additional information.

Earthquakes and tectonic activity

(a) In August 1999 an earthquake, measuring 7.4 on the Richter scale, hit Izmit in Turkey, killing an estimated 15,000 people. The map below shows the location of the earthquake in Turkey.

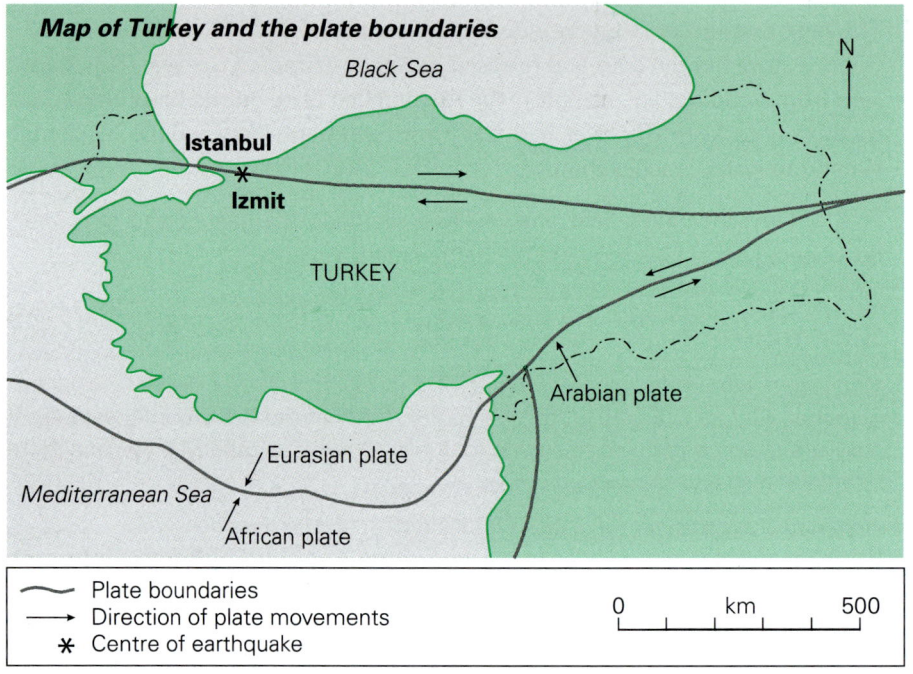

Map of Turkey and the plate boundaries
Black Sea
Istanbul
Izmit
TURKEY
Arabian plate
Eurasian plate
Mediterranean Sea
African plate
N

Plate boundaries
Direction of plate movements
* Centre of earthquake

0   km   500

(i)   What is an earthquake?   (1 mark)
(ii) Explain why Turkey is a high-risk area for earthquakes.   (4 marks)
(iii) Why could the number killed in the Turkish earthquake only be
      estimated and not known exactly?   (2 marks)

(b) Information about two earthquakes is given in Table 1 below.

Date	Place	Richter scale	Estimated number killed
1993	India	6.4	7,600
1994	California, USA	6.8	60

Table 1

(i) Describe what the information shows about the relationship between the strength of the earthquake and the number of people killed. (2 marks)

(ii) State **two** methods of protection to reduce the number of people killed in an earthquake. (4 marks)

(c) For **one** named earthquake, describe its effects upon the people and economy of the local area. (7 marks)

Total: 20 marks

Answer to Question 2: candidate A

(a) (i) The ground moves and buildings shake and collapse.

e The correct answer. It is only a 1 mark question so nothing more is needed. 1 mark.

(ii) I can see from the map that Izmit is on a plate boundary. Movement of large plates causes earthquakes. When plates move, shock waves are formed. There are three plate boundaries on the map, so a lot of Turkey can be affected by earthquakes. The one that goes through Izmit is called a conservative plate boundary, because the two sides of the plate are sliding past each other in different directions. Plates move in a jerky manner and this causes the earthquake. The same happens in California along the San Andreas fault.

e A good answer to this question must be based on plate boundaries, where tectonic activity is concentrated. The candidate refers to plate boundaries throughout the answer, so it is all relevant. The map is used and there is a general comment about plates in different parts of Turkey, which fits the question well. One type of plate boundary is named, the conservative boundary. The reason that earthquakes form along this type of boundary is also explained. The final reference to the San Andreas fault is relevant, because the same type of plate boundary is found in California where there is an equally high risk of earthquakes. The answer would earn all 4 marks.

(iii) A lot of people were killed because Turkey is a poor country. The government did not have emergency teams ready to rescue people. This is why it was estimated that so many people were killed. This could be expected because it was a strong earthquake, 7.4 on the Richter scale.

e The candidate has not answered the question set. The question the candidate has answered is 'Why were so many people killed in the Turkish earthquake?' The question asked why it was impossible to find out the exact number of dead. Reasons that could have been given are the amount of destruction, bodies hidden under piles of rubble so that they were never found, and everything destroyed by fires. 0 marks.

(b) (i) The earthquake in California was stronger by 0.4 than the earthquake in India, but 7,540 more people were killed in India. This shows that there isn't a relationship between strength and number killed. This is because the USA is a rich MEDC whereas India is a poor LEDC and America can afford more protection.

e The first sentence is good enough for both marks. In fact, it is the best way of answering because, by calculating the differences between the two earthquakes, the candidate has used (and not just repeated) the values. The second sentence is useful for confirming the candidate's good understanding. The third sentence is not needed, because the candidate is beginning to explain, when the question only required candidates to describe. It would therefore be ignored by the examiner. 2 marks.

(ii) 1. Build earthquake-proof buildings. This is done by strengthening the structure with steel and also by paying special attention to the foundations so that the building can vibrate in an earthquake, e.g. using rubber foundations. In addition, keep buildings low to make them stronger.

2. Improve methods of prediction of earthquakes, then people can be evacuated before the earthquake happens. This can be done by asking scientists to set up more seismographs.

e Making buildings earthquake proof is one method of protection. Good detail was added by the candidate about how this can be done, so the first part of the answer would earn at least 2 marks. However, the second method stated is not acceptable and would earn no marks. Earthquakes cannot be predicted. Seismographs measure the strength, but only after the earthquake has happened. The candidate could have stated low buildings as the second method. At least 2 marks would be given for this answer.

(c) In San Francisco in 1985 there was an earthquake along the San Andreas fault reading 6.9 on the Richter scale. Almost 1,000 people were either killed or injured. Many people lost their homes in the Marina area of the city, because they had been built on reclaimed land. This meant that when the earthquake hit, the land just collapsed, taking the houses with it, because the houses were built on soft, wet land instead of hard rock. There was a fire which increased the amount of damage and the number left homeless. This was followed by a lack of safe drinking water, and clean water had to be brought in by tanker. Part of the Bay Bridge collapsed too. The earthquake happened at 5.09 p.m., which was the worst possible time because it was rush hour and the roads were busy. People were injured and trapped in their cars on the way home. Billions of dollars worth of homes were lost, but many of the new skyscrapers survived with little damage because they were built to be earthquake proof. The cost of damage to homes was estimated at over $4 billion.

e This is a very good answer about the effects of the San Francisco earthquake upon people. There is plenty of specific information about this named earthquake throughout the answer. But there is not enough about the local economy part of the question — this is only mentioned in the final sentence. The candidate could have made more references, for example, to the disruption to businesses caused by the collapse of one section of the Bay Bridge and the difficulties for people reaching their places of work. The many good points about the answer cannot fully offset the shortage of economic references. 6 marks.

Summary

What mark would the answer be given?
15 out of 20, possibly 16 if the examiner was prepared to be more generous with the answer to (b)(ii). It would be a borderline grade A, not a certain grade A.

How could the answer have been improved?
By checking that the question set had been fully answered. After finishing an answer, read the question again before moving on to the next part.
Check that:
- the question has not been misinterpreted as in (a)(iii)
- no aspect of the question has been left uncovered, as 'economy' was in part (c)

Answer to Question 2: candidate B

(a) (i) Shaking of the ground where plates move.

 A short but correct answer. 1 mark.

(ii) There are many plate boundaries and fault lines. Some are moving in opposite directions. The Eurasian plate is moving towards and bumping into the African plate. This means that there will a lot of movement and shaking of the ground along where the plates are found, e.g. at Izmit near Istanbul.

 The answer is all about plate boundaries and the candidate has used the map. However, the types of plate boundaries are not identified and named, nor is there any explanation of the causes of earthquakes along these boundaries. In other words, this answer is too imprecise to gain all the marks, even though the candidate shows basic understanding. 2 marks.

(iii) Everything can be totally destroyed in an earthquake. There is rubble all over the place. In the Kobe earthquake large areas were also destroyed by a great fire. There was no way rescue workers could count all the bodies.

 This is a much better answer — clear, precise and to the point. It shows that the candidate understands well. 2 marks.

(b) (i) When the earthquake is stronger, fewer people are killed, which is not what you would expect.

 The basic point is made, but it is not supported by use of values. 1 mark.

(ii) 1. By making buildings earthquake proof. The government should see that this is done.

2. By having earthquake drills so that people know what to do when the earthquake happens. Go outside. Do not use lifts etc.

 1 mark would be gained for method one and 2 marks for method two. Method two is stated more fully, although a candidate is never going to be given any credit for using 'etc.', which is usually taken as a sign that the candidate has run out of knowledge. For method one, nothing is stated about how buildings can be made earthquake proof, which is why it would only contribute 1 mark. 3 marks.

(c) In 1995 an earthquake struck Kobe, a city in Japan. It measured 7.2 on the Richter scale and lasted only 45 seconds. Buildings, bridges and roads fell down because they had not been built to withstand earthquakes. People got crushed and killed by falling houses. Many people lost their lives and even more were injured. Families were affected because they had lost not only their homes but relatives too. Communities had to be built up again from scratch. It was also bad for the economy of the local area.

 The strength of this answer is that it is about a named earthquake. Its weakness is that there is little information specific to the Kobe earthquake. Apart from the first two sentences, all the information could be applied to any earthquake, anywhere. The only reference to economy is at the end, but the candidate does nothing more than add the word 'bad' to what is already written in the question. Only 3 of the 7 marks would be earned.

Summary

What mark would the answer be given?
12 out of 20. Grade C.

How could the answer have been improved?
By using geographical information that was more precise, e.g. in part (c), and by answering the shorter questions as fully as possible. A better eye could have been kept on the number of marks for each part of the question.

Exam tips

1 After writing down an answer, check that you have answered the actual question set and not the one you thought was set on first reading.

2 When there are two parts to a question, e.g. two methods in (b)(ii), and people and economy in (c), include answers to both parts and make them as equal as possible.

Question 3

Physical geography

Coasts

(a) Study Figures 1 and 2 showing the cliffs at Beachy Head before and after the storm of January 1999.

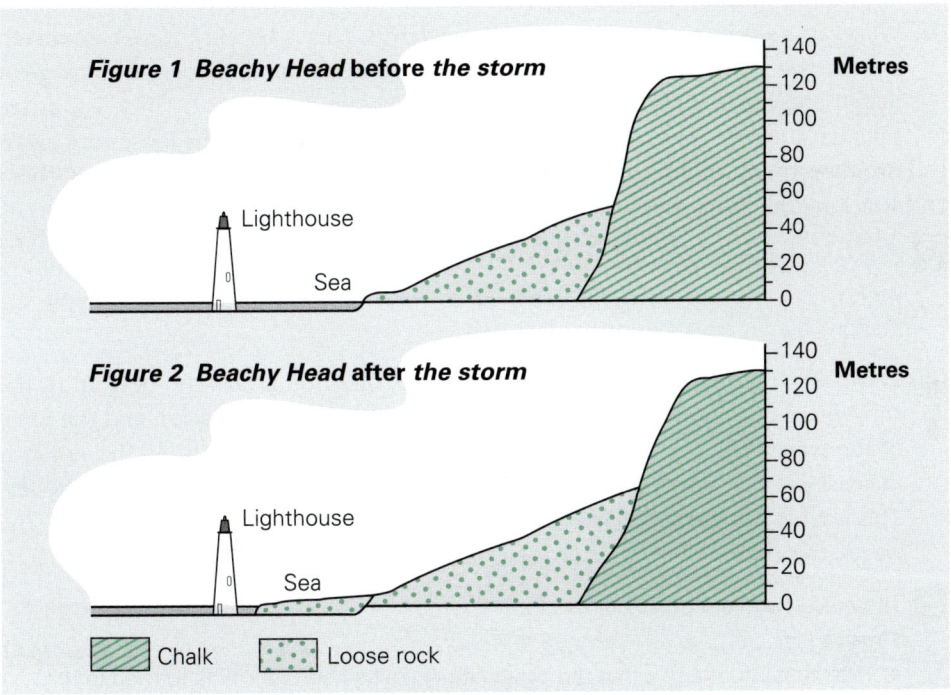

Figure 1 **Beachy Head** before *the storm*

Lighthouse

Sea

Figure 2 **Beachy Head** after *the storm*

Lighthouse

Sea

Chalk   Loose rock

(i)   Describe **three** differences after the storm. (3 marks)
(ii) Why do waves cause more coastal erosion during a storm than at other times? (3 marks)
(iii) Explain the formation of sea cliffs. (4 marks)

(b) Loose rock is picked up and transported along the coast by the long-shore drift.
(i)   Draw a labelled diagram to show how the longshore drift operates. (3 marks)
(ii) Name **one** landform that is formed by the longshore drift. (1 mark)

(c) The town council in a holiday resort wants to widen the beach and protect the coastline behind it.
(i)   Describe **two** methods the council could use. (4 marks)
(ii) What might stop the council from being successful? (2 marks)

Total: 20 marks

Answer to Question 3: candidate A

(a) (i) There is more loose rock at the bottom of the cliff. The second difference I notice is that the rock has gone further out into the sea and almost reached the lighthouse. The third difference is that the cliff has retreated; it is further back from the sea than it was before the storm.

e Three correct differences are stated very clearly. 3 marks.

(ii) Waves in a storm are bigger and more powerful. This is because waves are driven against the coast by strong winds. They crash down on the coast from a great height and break off pieces of the loose rock shown in Figure 2. These waves are called destructive waves. They have strong backwashes to take the loose rock away. Another thing about them is that they occur often, with about 15 of them attacking the coast every minute.

e This is a really good answer. The candidate explains what happens when storm waves hit a coast and knows all about the characteristics of destructive waves responsible for coastal erosion. 3 marks.

(iii) Waves like the ones described in the previous question attack the bottom of the cliff, where they erode a wave-cut notch by erosion such as abrasion and the force of the water. The top of the cliff overhangs the wave-cut notch. As the notch is increased in size, the overhang becomes bigger until the cliff collapses into the sea. This is what has happened in Figure 2. As soon as the loose rock is removed by the waves, the chalk cliffs will be eroded more.

e There are two very good points about this answer which mean it would gain all 4 marks:
- The most important is that the candidate explains how the cliff is formed in the order that it happens: wave attack at bottom of cliff; formation of wave-cut notch; cliff overhang and cliff collapse.
- The candidate shows good understanding by referring to types of erosion and to Figure 2. Without these references, the answer would still have gained full marks, but they have improved its quality significantly.

(b) (i) The longshore drift goes from left to right on the diagram below. Waves approach the beach at an angle, but go straight back down to the sea at right angles. In this way a pebble is moved along the beach.

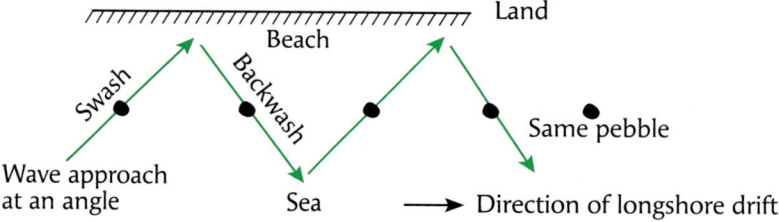

e The diagram is well labelled, but has been drawn carelessly. The backwash is *not* shown down the beach at right angles (90°) to the coastline. Although the candidate states this correctly in the written part, only what is labelled on the diagram will be marked in this question. 2 marks.

(ii) Spit.

e 1 mark. Other acceptable answers are beach, bar and tombolo, i.e. all the landforms of coastal deposition.

(c) (i) 1. Groynes. These are wooden boards built down the beach into the sea, which trap sand. They help to make the beach wider.
2. Wall. A sea wall with large blocks of rock in front of it will break the force of the waves. This will protect the coastline behind it.

e Two correct methods are named and described. By adding further details after naming each method, the candidate is obeying the command word 'describe' used in the question. 4 marks.

(ii) A big storm like the one at Beachy Head in 1999 might destroy the groynes and sea wall. Even if it didn't do this, it might take away all the sand off the beach. We cannot stop natural hazards doing damage.

e Good understanding is shown here. The sea is very powerful and can only be controlled for a short time by humans. 2 marks.

Summary

What mark would the answer be given?
19 out of 20. An A* answer.

How could the answer have been improved?
By taking more care drawing the diagram, to ensure that it was accurate in every respect.

Answer to Question 3: candidate B

(a) (i) 1. More loose rock.
2. Less sea.
3. Less cliff.

e These are short, simple answers. They do not describe as required by the question. Numbering the answers 1, 2 and 3 in questions asking for three items is fine, but you still need to describe by writing in sentences. Many examiners are likely to decide that there is no proper description in this answer and give it just 1 mark. 'Less cliff' is inferior to making the statement that the cliff has moved inland or retreated. 1 mark.

(ii) Strong winds whip up the waves in a storm. Big waves smash over sea walls and piers. Water pressure is much greater and this causes more erosion.

e The candidate gives part of the answer needed by making valid points about big waves and water pressure causing erosion. It is a pity that the answer does not go further, either by mentioning types of wave erosion that are increased, or by recognising that these waves are destructive waves. 2 marks.

(iii) Waves erode by corrosion, corrasion and abrasion. Cliffs collapse and the coastline goes further back. Soft rocks collapse faster than hard rocks. It is after a notch is formed at the bottom of the cliff by the waves, that the cliff collapses. An example of cliffs are the White Cliffs of Dover. These are made of chalk like the cliffs at Beachy Head.

e This is a chaotic answer. Types of wave erosion are named at the beginning, but they are not applied to cliff formation and are left isolated. Cliff formation is not explained in the order that it happens. The references to the formation of the notch and cliffs collapsing and retreating are the most relevant parts of the answer. 2 marks.

(b) (i)

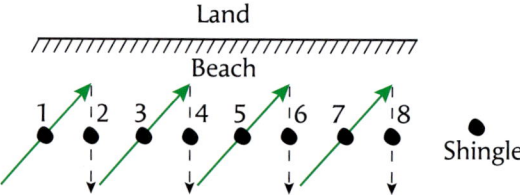

e The diagram is accurate and clear, but it is unfortunate that the direction of the longshore drift is not made clear and labels are in short supply. 1 mark.

(ii) Beach and spit.

e Both are correct, so the candidate gains 1 mark. However, if one had been wrong and the other correct, the candidate would have lost the mark. There is nothing to be gained from naming two landforms when one only is asked for.

(c) (i) 1. Wooden boards known as groynes.
2. Sea wall with an overhang at the top.

e Appropriate methods have been named, but not described. Only the minimum amount of additional information is given for each one and this is insufficient for the 2 marks for each method to be claimed. Marks would be earned only for naming each method. 2 marks.

(ii) Global warming.

e How? Global warming seems to be blamed for everything! It could have been made relevant by giving the context, i.e. because sea levels are rising and many believe that storms are becoming more frequent due to the effects of global warming. However, this is doing the work for the candidate. 0 marks.

Summary

What mark would the answer be given?
9 out of 20. Not enough marks to guarantee a grade C. It would only be given a grade D as it stands.

How could the answer have been improved?
By describing more fully and adding more information; by not using one or two word answers in questions worth more than 1 mark.

Exam tips

1 Be accurate when drawing labelled diagrams (and maps). You need to be aware that if something is wrong, it is often more obviously wrong on a diagram than when you are writing about it.

2 Except for short 1 mark answers, do not use one or two word answers or lists. Write in sentences so that your answer fits the context of the question. If you write several two-word answers, as candidate B did, your answers will not appear to be those of a candidate deserving grade C or above.

3 When explaining the formation of a landform, arrange the answer in the order of its formation.

Tropical rainforests

(a) Study Figure 1 below, a diagram of a tropical rainforest.

30 m

Figure 1

(i)   State **three** characteristics of the tall trees shown in Figure 1.   (3 marks)

(ii) What is the evidence for the existence of layers in this type of forest?   (2 marks)

(b) The climate of an area where tropical rainforest grows is given in Table 1.

Month	J	F	M	A	M	J	J	A	S	O	N	D	Total
Temperature (°C)	26	27	27	27	28	27	27	27	27	27	27	27	—
Precipitation (mm)	252	172	193	188	172	172	170	196	178	208	254	257	2,412

Table 1

(i)   Describe the main features of temperature and precipitation shown in Table 1.   (4 marks)

(ii) Explain **two** ways in which climate affects the characteristics of the tall trees in tropical rainforests.   (4 marks)

(c) Choose **one** area of the world where a large amount of tropical rainforest has been cleared.

    (i) Name and locate the area.   (1 mark)

    (ii) Give reasons for the clearance of the rainforests in the location named.   (6 marks)

Total: 20 marks

Answer to Question 4: candidate A

(a) (i) 1. They are 30 metres high.

           2. They only have branches at the top.

           3. They have buttress roots above the ground.

e Three correct characteristics are stated which would earn 3 marks. It is acceptable to answer by numbering 1–3 when three characteristics (or three reasons, ways etc.) are asked for in the question, but make sure that you give three that are accurate and different. If in doubt, answer in sentences.

(ii) The tall trees form a canopy, which is the top cover of the forest. On this diagram it is between 20 and 30 metres high. Another layer of not such tall trees is shown about 15 metres above the ground. They get less sunlight than the tall trees, which is why they don't grow as tall.

e Two layers of trees are identified by the candidate. Another good point about this answer is the way in which the candidate states heights, showing that the answer is based upon evidence from the diagram. The reason given in the third sentence goes beyond the needs of the question. 2 marks.

(b) (i) It is hot all year. In 10 months of the year the temperature is the same (27°C). The difference between the maximum temperature in May (28°C) and minimum temperature in January (26°C) is only 2°C. This gives a very low annual range of temperature.

           It is also wet all year. There is at least 170 mm of rain in every month and the total of precipitation for the year is high because it is nearly 2,500 mm. The months from November to January are the wettest with over 250 mm of rain per month.

e This can be used as a model answer for how to describe climate data.

- For temperature, quote the maximum and minimum temperatures and calculate the annual range. Use adjectives like 'hot' which describe the temperatures. Refer to months and temperature values, always quoting the unit of temperature (°C).
- For precipitation, describe the general pattern (in this example 'wet all year' and 'at least 170 mm in every month') and give examples of the distribution (in this example by referring to the wettest months). Refer to months and precipitation values, always quoting the unit (mm).

This answer would easily score all 4 marks.

(ii) 1. The trees are tall because they are trying to reach up high for the sunlight. They need the sunlight for photosynthesis. Only small trees can grow on the forest floor because of the darkness.

   2. The tall trees have buttress roots above the ground to give them strong supports, because they are so tall. They are also useful in obtaining nutrients because the top part of tropical soils is the only fertile part, as a result of rapid decomposition of dead vegetation in the hot and wet climate.

e This answer was almost a total disaster for the candidate because there was no mention of climate until right at the end. For point (1), the candidate gives a correct reason why trees are tall, but does not explain this in relation to climate so does not answer the question set. The climatic reason that the candidate could have used to explain the great height of trees is hot and wet weather all year round giving ideal conditions for tree growth. Under point (2), it appears that the answer is going to face the same problem, until the valid reference to climate at the end, which would allow the examiner to award 2 marks. The candidate has failed to use his/her knowledge of tropical rainforest vegetation to answer the question asked in an effective way.

(c) (i) My chosen area is the Amazon Basin in Brazil.

e 1 mark. The area has been named (Amazon Basin) and located (in Brazil). The candidate has responded to both command words in the question ('name' and 'locate').

(ii) One main reason why the Amazon rainforest has been cleared is for logging. Hardwood trees like mahogany have many uses and are exported, which makes a high amount of foreign exchange for Brazil. On the cleared land big companies can also set up cattle ranches and export meat, for which there is a great market in MEDCs because of the high demand for burgers.

The Amazon basin is rich in minerals such as iron ore, manganese, bauxite and gold. The main mining area is around Carajas. From here a railway line was built to transport minerals to the coast for export. Brazil is an LEDC and needs all the money it can earn from exporting timber and minerals to help with its economic development. The government wants Brazil to become a more economically developed country.

The government of Brazil has built roads to make it easier for people to settle in the interior and start farming. Examples of roads are the Trans-Amazonian and Marshal Rondon Highways. Poor people from the coast have migrated here, cleared the forest and started farming. In the east of Brazil there was no spare farmland for them, whereas in the Amazon land for farming is plentiful.

e This is a good case study answer, based upon the area named in part (i), because:

- several different reasons for clearance (logging, mining, farming and road building) are mentioned
- an example of a mining town is given and roads are named
- reference is made to economic development in Brazil

This answer falls into the top level and includes plenty of content specific to the case study. 6 marks.

 **Summary**

What mark would the answer be given?
18 out of 20. This would receive a grade A*.

How could the answer have been improved?
By double checking every answer to make sure that what was written matched the question set. In answering part (b)(ii) the candidate underestimated the importance of climate in the question and would have dropped 4 marks instead of 2 without the final reference to climate. Although the really good previous answer to part (b)(i) was worth more than the 4 marks, marks can never be transferred between different parts of a question. Once marks are lost, they cannot be recovered.

Answer to Question 4: candidate B

(a) (i) They are tall to reach the sunlight. All the branches are at the top. From the branches at the top, creepers hang down and some of them reach down as far as the ground below.

e No mark would be given for what is written in the first sentence; this is more like a reason than a statement and there is no indication of tree height. However, there would be 1 mark for each of the next two sentences, making 2 marks overall. It is good policy to state an extra characteristic when known, in case the examiner does not accept one of the answers, as happened here.

(ii) The canopy of tall trees forms one layer at the top of the forest. Another layer is formed by the vegetation on the ground and the tree roots. There are other layers of smaller trees between 10 and 20 metres above the ground.

e The candidate understands what is meant by layers and begins with a correct statement in the first sentence. However, no marks would be awarded for the second sentence; little vegetation is shown on the forest floor in Figure 1 and roots are not part of the forest layers. In the third sentence further correct information is given, which means this answer would receive 2 marks.

(b) (i) The only temperatures are 26, 27 and 28. This means that the temperature is almost the same all year. There is little change between winter and summer.

The precipitation begins at 252 in January, falls to 172 in February and goes up in March and April before going down again during May, June and July. December is the month with the highest rainfall. There seems to be quite a lot of rainfall in every month with a total of 2,412.

e Unlike candidate A, this candidate does not know what he/she is looking for to describe the key features of temperature and precipitation data. This candidate's approach to answering is to repeat data without quoting the units (°C and mm), often month by month, instead of picking out and describing the data that are significant. The answer would just about get 2 marks: 1 mark for temperature being the same all year and 1 mark for the general statement about rain in all the months with most in December. No marks would be given for quoting rainfall values month by month.

(ii) Trees grow tall in tropical rainforests to reach the sunlight because it is hot and wet all year. They have large leaves with drip tips at the end of them so that the heavy rainfall drops off them during the downpours, which happen on most days of the year. Creepers use the trees as support as the top of the trees is a good place for them to grow because of the sunlight, heat and high rainfall.

e This is a better answer than the one given by candidate A, because something related to the effects of climate is included in every statement about the tall trees. It would earn all 4 marks: 2 for the first sentence and 2 for the second sentence. A third way is also covered; as stated earlier, it is a good idea to include one, but it is not needed on this occasion.

(c) (i) Brazil.

e By only naming a country the candidate does not give an answer that is sufficiently precise to claim this mark. At least it is enough to suggest the location that is going to be used in the next part of the answer. 0 marks.

(ii) The forests have been cleared from the Amazon Basin and the land is now used for farming crops, cattle ranching, mining and roads. Houses and settlements are built on the cleared land. The wood is sold to other countries, where it is made into furniture and paper. In some places trees are replanted with seeds. The government of Brazil is making a lot of money by exporting logs, minerals and beef from the Amazon Basin.

But the clearance of the forests has caused many environmental problems. The natural environment changes because habitats are destroyed, more carbon dioxide goes into the air so that global warming gets worse and there is less tree cover, so that when it rains more flooding occurs and soil is washed away. This is bad for Brazil and for people living everywhere else in the world.

e Both the Amazon Basin and Brazil are named in this part. If both had been stated in (c)(i) they would have gained 1 mark. As marks are not transferred between questions, the candidate missed out on the mark.

The second paragraph is irrelevant because the candidate refers to consequences rather than to causes of forest clearance. No marks are deducted for writing down information that is irrelevant, but no marks would be gained either.

The first paragraph of this answer is the only one which would earn any marks. The first sentence is a list of land uses rather than reasons and the answer is short of reasons throughout. Overall, this is a low-level answer which would only earn 2 marks.

Summary

What mark would the answer be given?
12 out of 20. This would be a grade C. It is generally well answered in parts (a) and (b), but not as well in part (c).

How could the answer have been improved?
- By giving a more precise and detailed answer to the case study part of the question.
- By concentrating upon reasons instead of consequences in the final part.
- By being more aware of what to look for when asked to describe climatic data.

Exam tips

1 Do not ignore key words in questions such as 'climate' in (b)(ii) and 'name and locate' in (c)(i).

2 When asked to describe climate data (irrespective of topic):
- for temperature quote maximum temperature, minimum temperature and annual range supported by references to months and values with units
- for precipitation describe the annual pattern of wet and dry periods (not month by month) supported by references to seasons and values with units

3 Avoid writing sections in answers that are irrelevant, such as *consequences* of forest clearance when *reasons* for clearance are asked for.

World population growth

(a) Study Figure 1 which shows total world population over 200 years (from 1830 and predicted until 2030).

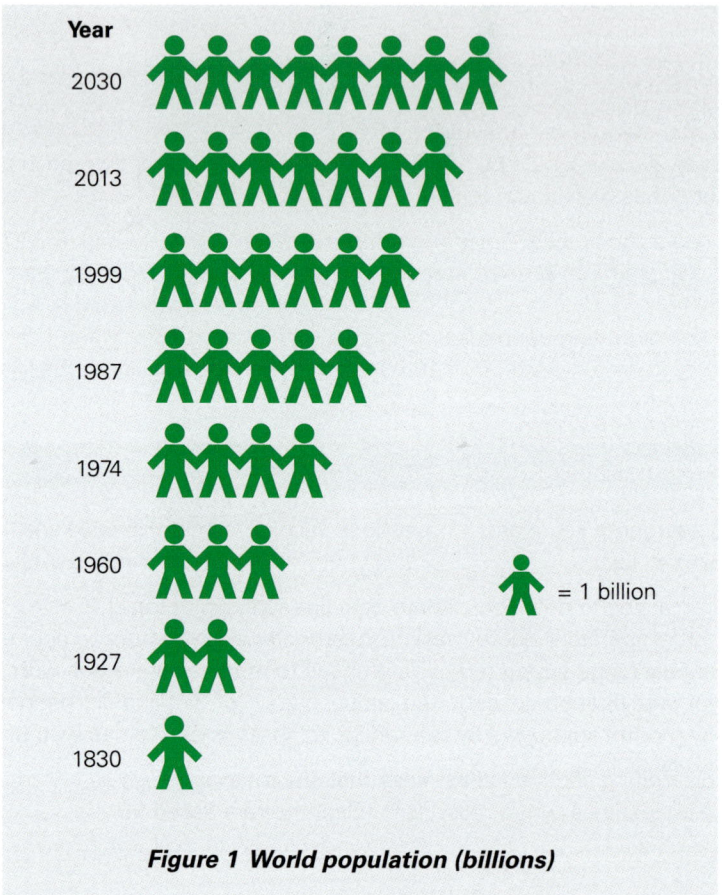

Figure 1 World population (billions)

(i)   How many years did it take for world population to double from 1 to 2 billion?   (1 mark)

(ii) Between which two years did world population increase by 1 billion in the shortest length of time?   (1 mark)

(iii) Describe the evidence from Figure 1 that the speed of world population growth is predicted to slow down between 1999 and 2030.   (2 marks)

(iv) Give **two** reasons why world population growth is expected to slow down in future.   (4 marks)

(b) Table 1 gives information on population for six continents.

Continent	Birth rate per 1,000	Death rate per 1,000
Less economically developed:		
Africa	43	14
Asia	26	9
South America	26	7
More economically developed:		
Australasia	20	8
Europe	12	10
North America	16	8

Table 1

Using the information in Table 1, name the continent with:
(i) the largest natural increase
(ii) the smallest natural increase
(iii) a natural increase of 12 per 1,000   (3 marks)

(c) Explain why birth rates remain high (above 25 per 1,000) in many
LEDCs.   (7 marks)

(d) Table 1 shows that death rates in Europe are on average higher than
those in Asia and South America. Suggest **one** reason for this.   (2 marks)

Total: 20 marks

Answer to Question 5: candidate A

(a) (i) 97 years.

e 1 mark.

(ii) 12 years between 1987 and 1999.

e 1 mark for stating the correct dates.

(iii) The evidence is that it is predicted to take 14 years for the population to increase
from 6 to 7 billion and then 17 years for it to increase the next billion.

e The calculations are correct and the candidate states relevant evidence.
However, the answer has not been finished off because the candidate makes
no attempt to describe how the stated evidence indicates slow-down in
population growth. You cannot expect an examiner to do the work for you.
1 mark.

(iv) 1. Birth rates are going down as more and more people are using condoms and other methods of birth control, e.g. in Brazil where the government runs clinics in rural areas.

2. AIDS is killing thousands of people in Africa. So many people are already infected with HIV that death rates in many African countries, e.g. Botswana, are increasing and will continue to increase.

e Two reasons that are clearly different, one for birth rates and one for death rates, are stated in a precise manner. Mentioning places is another good feature of this answer, greatly increasing its geographical worth. 4 marks.

(b) (i) Africa.

(ii) Europe.

(iii) Australasia.

e When the candidate, as here, understands how to work out natural increase (birth rate minus death rate), these are an easy 3 marks.

(c) There are still many reasons why people in LEDCs have large numbers of children. People in some countries like having large families, because everyone comes from a large family themselves.

1. Lack of family planning. Some governments, e.g. Somalia and Ethiopia in Africa, are too poor to set up family planning clinics and provide contraceptives to people in rural areas.

2. High death rates of babies. Many children die before reaching their first birthday; others die before they reach 5 years old, so there is a lot of infant and child mortality. Mothers try to compensate for these high death rates by having many children so that at least some of them will live until the age when they can start working and earn money.

3. Children are seen by many people as assets. Children may not go to school because they are expected to work. Boys help with farming from about the age of 9 or 10 and they make money when their parents become old.

For these reasons birth rates are high in many LEDCs.

e Three reasons are used to explain the high birth rates. The candidate elaborates with a reasonable amount of detail and an occasional example. Is there sufficient content for all 7 marks? Other reasons could have been given, such as the lack of education of women and religious objections to the use of birth control, but in an answer at GCSE level no candidate can be expected to include all possible reasons. This answer would receive at least 6 and possibly all 7 marks.

A note about numbering answers — in an earlier question, part (a)(iv), two reasons were asked for, which was an open invitation to the candidate to number the reasons 1 and 2. However, numbering reasons, and taking them one at a time, is not the recommended way of answering longer GCSE questions, because an answer can be more effectively communicated with continuous writing. When it is a close decision whether an answer is worth 6 or 7 marks, the candidate who numbers the reasons is likely to lose out and be given only 6 marks.

(d) This is because medical services are much better in MEDCs than in LEDCs. There are more doctors per head of the population and governments can afford drugs to keep people alive, whereas many people are too poor in LEDCs and live in the countryside where there are no hospitals.

 0 marks because the candidate is not answering the question. A higher proportion of old people live in Europe than in Asia and South America. Although medical services are better in Europe, the presence of a higher percentage of old people means that more die per 1,000 of the population each year, simply because of old age. In LEDCs there is a much higher proportion of young people, which helps to keep the death rate per 1,000 people low.

Summary

What mark would the answer be given?
16 out of 20. Grade A. Few mistakes were made and only the difficult question at the end was wrongly answered.

How could the answer have been improved?
By writing in continuous English in long answers rather than numbering points.

Answer to Question 5: candidate B

(a) (i) I can see that it is about 100 years.

 0 marks. It is possible to give an accurate answer from Figure 1 (97 years), so only this answer will be accepted.

(ii) I think this was between 1987 and 1999 because the difference of 12 years was the smallest.

 1 mark for stating the correct dates. (As in candidate A's answer, the number of years was not needed, but no harm was done by stating this.)

(iii) I can see from the diagram that it will take 31 years for world population to go up from 6 to 8 billion from 1999 to 2030. I can see that it took only 25 years for the world population to go up from 4 to 6 billion from 1974 to 1999, which is less time for the same amount of growth in population. This tells me that population growth is slowing down after 1999.

e This candidate may not have taken the fastest and easiest route to answering, and expression is clumsy, but he/she does answer the question set, which is what matters most. 2 marks.

(iv) 1. I think that birth rates will fall because more people will use family planning and attend family planning clinics.
2. I know that some governments in LEDCs are giving free condoms to young married couples so that they will have fewer children than their parents had.

e The main problem with this answer is that the two reasons numbered 1 and 2 by the candidate are too similar. In effect, the reason is the same — greater use of family planning is slowing down population growth. For marking purposes, the candidate gives one reason, which is quite well developed, and this means the answer would be worth 2 marks.

(b) (i) Africa.

(ii) Europe.

(iii) Europe.

e Only the first two answers are correct. Actually the candidate may have been a little fortunate here, because the wrong answer of Europe to part (iii) suggests that the candidate was using birth rate instead of natural increase throughout. 2 marks.

(c) 1. No birth control used.
2. No family planning.
3. Religion, e.g. Catholics are told not to use birth control.
4. High death rates.
5. Children help with work on the farm.
6. Look after parents in old age.
7. Governments are too poor to educate people about family planning.

e This is definitely *not* the way to lay out an answer to a 7-mark question. Giving headings does not allow anything more than superficial explanation.

Trying to take a positive view of this mini-disaster in style of answering, at least the candidate did list seven reasons, which matched the number of marks for the question. The reasons listed are valid, even if there is some overlap. For example, numbers 1 and 2 are the same. This also happened in the earlier answer to (a)(iv). Finally, an example is given to illustrate reason number 3.

How would it be marked? Making a list like this keeps the answer in the first band of assessment. Often a maximum of 2 marks will be given to a list-like answer. Given the length and relevance of the list as well as mention of an example, the most likely mark would be 3.

(d) I think this is because medical care is much better in MEDCs than in LEDCs.

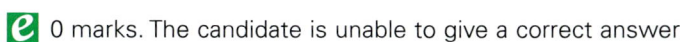

 0 marks. The candidate is unable to give a correct answer.

Summary

What mark would the answer be given?
10 out of 20. Grade D.

How could the answer have been improved?
Mainly by changing the way in which the answers were written. The comment in part (c) about the need to avoid numbering and listing in long answers is of critical importance. However, there is another expression issue here. Did you notice how this candidate begins the majority of answers with 'I think', 'I know', 'I can see'. Using this personalised style wastes time and words, and tends to give an examiner the impression that the answers are simple and flimsy. Using 'I' should in general be avoided in answering examination questions, except those which ask for your views on a subject.

Exam tips

1 When asked for two or more reasons, try to ensure that they really are different and not just the same one restated in another way.

2 Do not use numbered lists to answer longer questions. The only time it is safe to use them is in answering a question that asks for a specific number of reasons or ways.

Pyramids and population structure

(a) Study the population pyramids for Ethiopia and the UK.

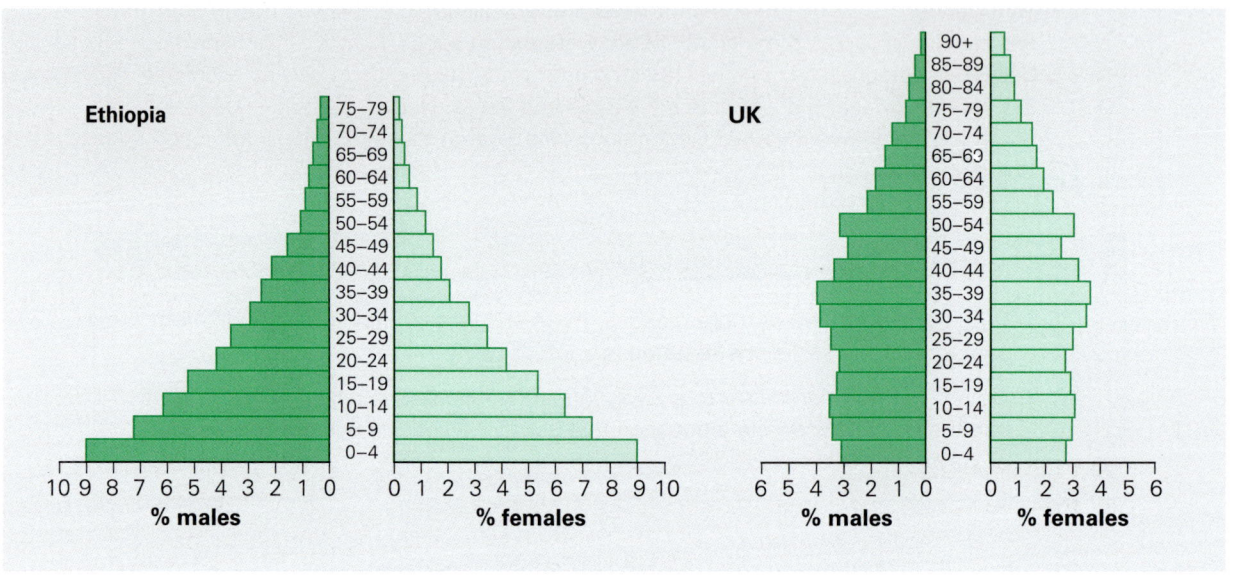

(i) What percentage of the population of Ethiopia is aged 14 years and under? (1 mark)

(ii) Describe **one** difference in shape between the two pyramids. (2 marks)

(iii) How do differences in birth rate and life expectancy affect the shape of the two pyramids? (2 marks)

(b) For **one** named country, describe the methods used to reduce birth rates. (5 marks)

(c) The population of a country can be split into three groups:

A: 0–14 years: young dependants

B: 15–64 years

C: 65 and over: old dependants

(i) Give a label suitable for the group aged 15–64 years. (1 mark)

(ii) Explain why the other two groups are called 'dependants'. (2 marks)

(d) The UK and many other MEDCs have ageing populations (increased percentage of old people). Describe fully the consequences for a country of an ageing population. (7 marks)

Total: 20 marks

Answer to Question 6: candidate A

(a) (i) 46%

> *e* The answer is comfortably in the middle of the percentage range that would be allowed in the mark scheme. 1 mark.

(ii) The shape for Ethiopia looks like a real pyramid with a wide base and steps up to the narrow top. The UK's pyramid is almost straight up and down.

> *e* A positive, separate statement is made about each graph. One difference is stated clearly. 2 marks.

(iii) Ethiopia has a much higher birth rate than the UK and so its graph is much wider at the bottom. Ethiopia has a much lower life expectancy than the UK, so that its pyramid is narrower and less high at the top.

> *e* Well answered again. Differences are explained very clearly in relation to birth rate and life expectancy. 2 marks.

(b) China's population is almost a quarter of the total world population. With worries that China would become overpopulated, a one child policy was introduced. Each couple was only allowed one child. If you wanted a child you would have to receive a permission card. Without this you would be unable to register the child. Couples with one child would receive a 10% increase in salary; however, couples with more than one child would receive a 10% decrease. The government provided family planning advice and clinics and had massive advertising campaigns to get people to stop at one child.

> *e* The answer is entirely about the candidate's named country, China. The candidate concentrates upon describing methods used to limit birth rates, all of which apply to China. The description is full enough to earn all 5 marks.

(c) (i) Workers. Old and young depend upon the money made by them.

> *e* The proper name for this group is 'economically active'. A simpler label is 'people of working age'; therefore, workers is an answer which most examiners would be willing to accept, even if it is not the best. 1 mark.

(ii) Old and young depend upon money made by workers. The money workers pay in taxes is used by the government for schools for the under-15s and for pensions for the over-65s. Old and young do not earn money which is why they depend upon the workers.

 A dependant is someone who needs to rely upon another person. This candidate makes this clear and understands perfectly what dependency means. 2 marks.

(d) As a result of improved medical knowledge, more diseases and illnesses can be cured. However, as people get older and live longer they need more and more medical help. This puts a great strain on the NHS (National Health Service). Some cannot look after themselves any more either, and have to go into care homes run and paid for by local councils. People over 65 do not work, so they are no longer paying into the economy; however, they are taking money out with pensions, which they need to live on. Pensions are paid out by the government using the money it has collected in taxes from those who are younger and still in work. As the percentage of old people increases, the amount the government needs to pay out increases. Some governments in MEDCs are now worried that they will not have enough money to keep paying pensions to all of their old people, especially if the workforce gets smaller. In addition, raising taxes from those still in work is not going to be popular.

 Several consequences are described — for the health service, care homes and pensions. All are valid and placed in the right context, because many governments are worried about how they will be able to raise the funds needed in future to look after an increasing number of old people. If there is a criticism of this answer, it is that the candidate has only dealt with the 'bad' consequences, or disadvantages, of an ageing population. Are there any 'good' consequences or benefits? Yes. For example, the tourist industry and companies like Saga benefit as old people have more free time and can take their holidays during less busy times outside school holidays. 6 marks.

Summary

What mark would the answer be given?
At least 19 out of 20. This is right at the top of the grade-A band. A very good answer like this gives the candidate a few spare marks, in case one or more of the other answers are not as good, which increases the chance of gaining an A or A* grade overall in GCSE geography.

How could the answer have been improved?
Only by looking for both good and bad in a question that asks for 'consequences'. Apart from this, the answers are as good as can reasonably be expected from a GCSE candidate.

Answer to Question 6: candidate B

(a) (i) 24%

 0 marks. It is likely that the candidate added up one side only, and gave the total for either males or females, but not for both males and females as was needed.

(ii) The population pyramid for Ethiopia has a wide base; the UK's hasn't. This is because there is a much higher birth rate in Ethiopia.

e This answer would only be given 1 mark because only one side of the difference is directly stated. Nothing is said about the UK pyramid that is positive. Mentioning higher birth rate does not help to claim the second mark, because it is explanation that is irrelevant in this part of the question. Did the candidate read ahead to the next part of the question?

(iii) The much higher birth rate in Ethiopia gives a wide base to its pyramid. The higher the life expectancy, the wider the top, which is what I can see on the UK pyramid.

e The first sentence, which is almost a repeat of the answer from the previous part, is relevant in this part and would gain the first mark. The second sentence about life expectancy is also true. 2 marks.

(b) China. The one-child policy. This meant that people did not receive benefits for a second child.

e A correct answer, but much too brief for a 5-mark question. Extra details about methods used, like those given by candidate A, are needed. 2 marks.

(c) (i) Working population.

e This is an acceptable answer. 1 mark.

(ii) 'Dependant population' means the section of a population that doesn't work, whether they are too young or too old. Old dependants are retired people on pensions. Young dependants are those at school and too young to work.

e The candidate does quite a lot of repeating of young and old dependants from the question, without adding much extra detail. Mention of pensions would earn 1 mark with reference to old dependants. However, full understanding does not come across in reading this answer.

(d) • More pensions. Longer life expectancy and the government has to pay more people for longer.
• More health care. Drugs are keeping more old people alive.
• More care homes. These are needed for very old people. It costs a lot of money to run a home for old people because nurses are needed.
• Many old people go on cruises and take lots of holidays, which is why holiday companies like old people.
• Makers of zimmer frames and stair lifts make more money.

e In one way, this is quite a reasonable answer, because the candidate recognises both unfavourable and favourable consequences of an ageing population. At least four different consequences are identified. The 'More care homes' bullet

point is the most successful one because the candidate tries to give further information about why care homes are expensive to run.

In another way, it is a disappointing answer, because using bullet points is not the most effective way of conveying information in a long answer. It doesn't read as well as continuous writing. Some of the written expression is weak. Count how many times 'more' is used.

Depending upon the examiner's attitude to the use of bullet points, this answer would be given 4 or possibly 5 marks for its content, but the candidate is taking a big risk using bullet points in a longer answer.

 Summary

What mark would the answer be given?
11 (or 12) out of 20. Although there is variation in performance levels between the answers to different parts of the question, this would be a grade C overall.

How could the answer have been improved?
By studying more carefully what was needed in the short questions and by not using bullet points in the long answer.

 Exam tips

1 Do not use bullet points in long answers. Points can be scored more clearly and effectively by writing in sentences and paragraphs.

2 Before answering one part, look ahead to the next part of the same question. This reduces the risk of giving the same answer to two different parts — when you do this, one of the answers will be irrelevant.

Distribution and density

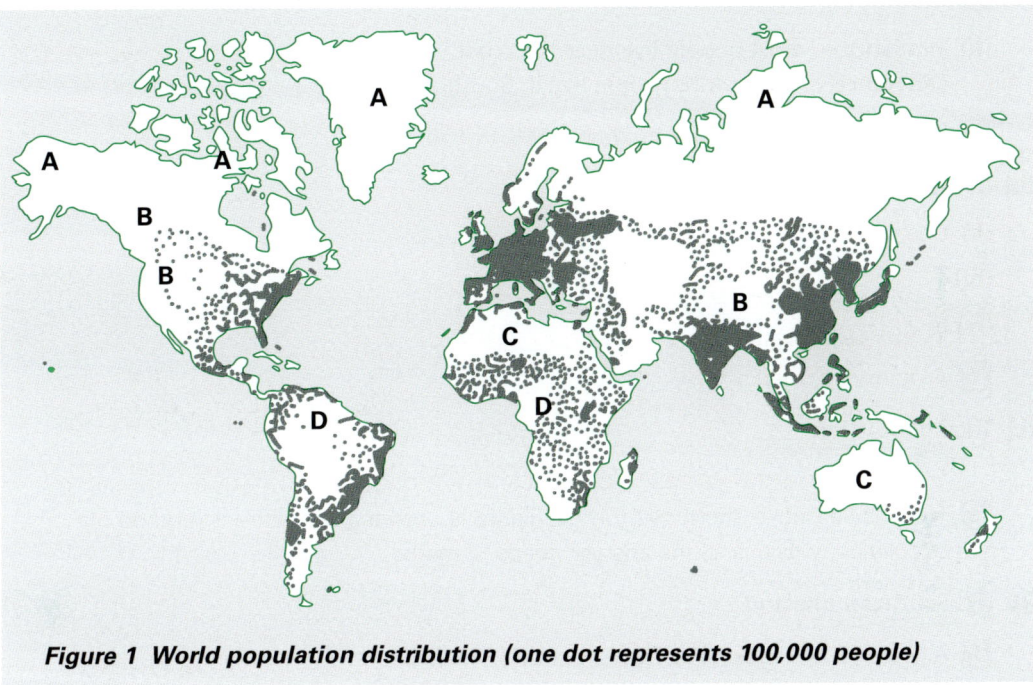

Figure 1 World population distribution (one dot represents 100,000 people)

(a) (i) Name the **two** continents with the largest concentrations of people in Figure 1. (2 marks)

(ii) State **one** similarity and **one** difference in population between Australia and South America shown in Figure 1. (2 marks)

(b) Very few people live in those areas of the world lettered A–D in Figure 1. Give the letter for the areas where the low number of people living can be explained by conditions being:
(i) too cold
(ii) too hot and wet
(iii) too dry
(iv) too high and steep (2 marks)

(c) How is the density of population in an area calculated? (2 marks)

(d) (i) Name **one** area you have studied with a high density of population. (1 mark)
(ii) Give reasons why so many people live in the area you have named. (6 marks)

Total: 15 marks

Answer to Question 7: candidate A

(a) (i) Europe and South Asia.

e 2 marks. Asia would have been acceptable; South Asia is a more precise answer.

(ii) Similarity — most people live near the coast.
Difference — a lot more people live in South America.

e Both statements are correct. 2 marks.

(b) (i) A

(ii) D

(iii) C

(iv) B

e All correct. 2 marks.

(c) $\dfrac{\text{Total population}}{\text{Area}}$

e The answer looks short, but the candidate is showing the correct method of calculation, which is all the answer needs. 2 marks.

(d) (i) Southeast England.

e A good choice. Looking at Figure 1 shows that it is an area with a great concentration of people. 1 mark.

(ii) • London is the capital city.
  • It attracts workers from the north of England where mines and factories have closed.
  • Many jobs are found in London, e.g. in shops, offices, transport, tourism and in government.
  • London is the largest financial centre in Europe.
  • It has airports, such as London and Gatwick, and attracts many visitors from overseas.
  • The southeast is the nearest part of the UK to Europe and other countries in the EU.
  • The main market for manufactured goods is in Europe.
  • Hi-tech industries, such as computers and electronic companies, like to locate in the southeast, e.g. along the M4 corridor between London and Reading.

All these reasons explain why many people live in southeast England and the numbers are going up all the time as more and more people move from Scotland and the north of England, places without enough jobs where the climate is colder than in the southeast.

ℯ This long answer contains more than enough geographical information for all 6 marks. The answer is mainly based upon the availability of work, which is an important reason why so many people live in the southeast. One good point is the frequent use of specific information, naming types of work and places. The candidate has avoided a common weakness of answers about density of population, namely lack of real information about places.

The main problem with this answer is the bullet point layout, which is not recommended for longer answers. Notice how much better the answer reads at the end when the candidate switches to writing in sentences. The candidate is likely to be given 5 instead of the total of 6 marks, because bullet point answers cannot meet all the requirements for effective written communication.

Summary

What mark would the answer be given?
At least 14 out of 15. This is an A*-standard answer.

How could the answer have been improved?
The content does not need improvement. However, although it is sometimes acceptable to use bullet points in short 2- or 3-mark answers, they are not an effective way to answer a 6-mark question requiring reasons.

Answer to Question 7: candidate B

(a) (i) 1. The UK and Europe.
2. India and China.

ℯ The candidate is naming countries instead of continents, but would gain 1 mark for Europe.

(ii) 1. Both have large areas with no population.
2. More people live down the east coast of South America.

ℯ 2 marks.

(b) (i) A

(ii) B

(iii) C

(iv) D

ℯ 1 mark would be given for selecting A and C correctly. However, B and D have been placed incorrectly. The high mountains of the Himalayas and Rockies, labelled B, have been interchanged with the hot wet tropical areas in the Amazon and Congo Basins, labelled D.

(c) Density is the amount of people per square kilometre, e.g. 1,000 people per km² in an area.

e The candidate understands what the density of population is, but fails to make clear that both total population and size of area are needed for its calculation. Only 1 mark. Saying 'number' of people would have been more accurate than 'amount' (which suggests that people are being weighed rather than counted!).

(d) (i) São Paulo in Brazil.

e 1 mark. It is a large, densely populated city. But is a city the best choice as a basis for the answer to the next part?

(ii) Many people have moved from the northeast of Brazil into São Paulo. Two thousand people are moving into the city every week. This is a result of rural to urban migration. Most people are attracted by jobs and the bright lights of the big city. Many of the migrants live in shanty towns until they find work. Shanty towns are found all around the edges of São Paulo on unused land where people build shacks and homes from any kind of materials they can find. In the northeast of Brazil the land is dry and people are poor, which is why they migrate to big cities in the hope of getting a job and having a better life.

e The main problem with this answer is that the candidate spends too much time writing about rural to urban migration and its causes. Obviously, such migration has contributed to the large number of people living in the city, but there are other reasons for its growth. The most important is that São Paulo is the largest centre of manufacturing industry in Brazil (and indeed in the whole of South America), about which there is no mention in this candidate's answer. There are too many non-specific references to jobs. Overall, the answer is short of real information about São Paulo. For this type of question it is often better to choose as large an area as possible to write about, giving you more reasons to explain why many people live there. For example, if southeast Brazil had been the named area, reasons which applied to Rio de Janeiro and the rich farming areas in between the cities could have been used as well. This answer would only score 2 marks because of its narrow and imprecise coverage.

Summary

What mark would the answer be given?
8 out of 15. This would probably be a grade C, but only because the short opening questions based on Figure 1 were relatively straightforward and allowed easy marks to be claimed.

How could the answer have been improved?
By making reference to a wider range of reasons and giving more detail about the named location in (d)(ii), essential in a 6-mark answer based on a case study.

Exam tips

1 By all means use bullet points in the plan for an answer, but do not use them in the answers to questions that are worth 4 or more marks.

2 Examination answers from candidates on density of population are notorious for the shortage of real detail about areas and places. Try to include as much specific information as possible, thereby making your answers better than those of other candidates.

Urban growth in LEDCs

(a) Study Figure 1 which gives information about population in Lima, the capital city of Peru in South America.

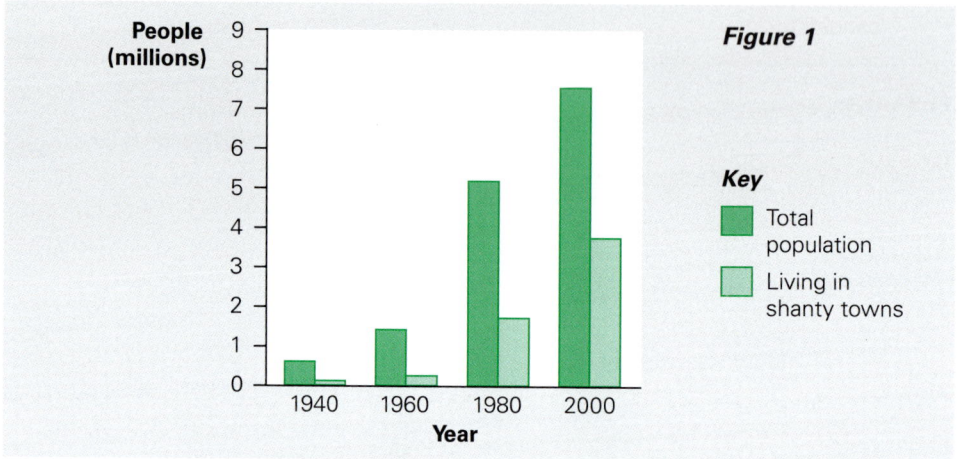

Figure 1

People (millions)

Key
■ Total population
■ Living in shanty towns

Year

Describe what Figure 1 shows about:
(i)   population growth   (1 mark)
(ii) numbers living in shanty towns   (2 marks)

(b) Give **two** reasons for the rapid growth of big cities in LEDCs such as Lima.   (4 marks)

(c) Look at Figure 2, which is a sketch of a house built by a family which has recently arrived in Lima.

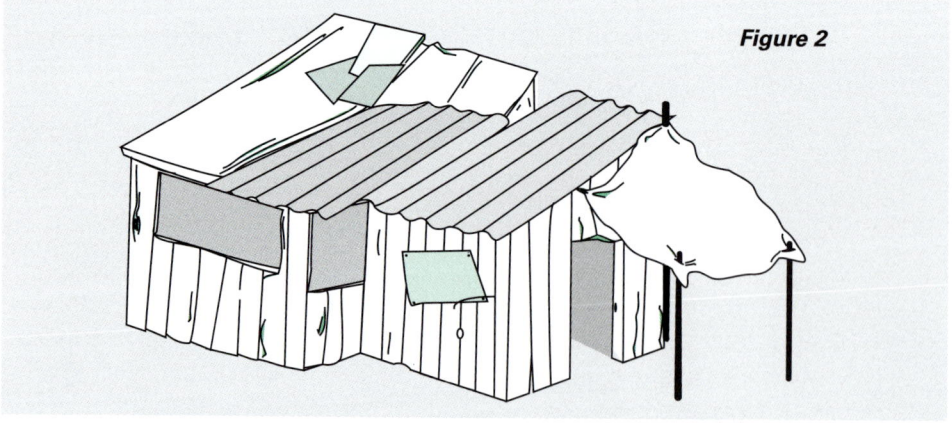

Figure 2

(i) What suggests that it was built by newcomers to the city? (2 marks)
(ii) Where in the city is it most likely to be located? Explain your answer. (2 marks)
(iii) What changes might be made to the house in future years? (2 marks)

(d) Describe the consequences, other than for housing, of the rapid growth of big cities in LEDCs. (7 marks)

Total: 20 marks

Answer to Question 8: candidate A

(a) (i) From a population of just over half a million in 1940 Lima's population has grown to $7\frac{1}{2}$ million in 2000, a 15 times increase.

1 mark. The key to picking up the mark is the use of two values; noting that it was a 15 times increase is extra, but valuable in suggesting to the examiner that he/she is marking the work of an able candidate. First impressions are important.

(ii) From hardly any in 1940 (too small to read off the graph), numbers living in shanty towns have gone up to almost 4 million in 2000. Another thing that has happened is that the numbers living in shanty towns make up an increasingly large percentage of the total until by 2000 they make up about half of the total population.

2 marks. More good understanding is shown. For this question there is 1 mark for increased numbers and 1 mark for shanty town dwellers making up an increased proportion of the total. This candidate claims both marks, again helped by the use of values from the graph.

(b) One reason and the main one in my opinion is rural to urban migration. This is happening in São Paulo in Brazil where people are leaving the northeast because of drought (a push factor) and moving to the southeast where big cities with industries and jobs are found and people feel that they will have a better standard of living (a pull factor).

The second reason is high birth rates. Women often have five or more children. There are lots of young people looking for work in big cities and they still want large families when they get married. Children are often a source of income, selling and begging on the streets.

Two totally different reasons are given, which is ideal. Both are important reasons too — in fact, they are the two most important reasons for big city growth in LEDCs. The candidate elaborates quite fully and makes certain that he/she gains the second mark for each reason, without writing too much and using time needed for later questions. A very comfortable 4 marks.

(c) (i) It looks self-built and is made of many materials. Newcomers to the city have no money and cannot afford to buy or rent a proper house, so they have to go and collect whatever they can and find some unoccupied land to build it on.

e The candidate shows good understanding. 2 marks.

(ii) They locate wherever unoccupied land exists in the city. They cannot afford to buy land so they have no choice where they place the house. Some build it next to where friends and relatives already have houses.

e The weakness of this answer is that the candidate does not suggest an actual location in the city. The same understanding shown in earlier questions is present again, but the 'where' part of the question has not been fully answered and therefore the answer would only be worth 1 mark. Possible locations that the candidate could have picked include the edges of the city, along the sides of rivers, and land too steep or too swampy for permanent houses.

(iii) Most people expect to find a job and then earn enough money to buy building materials. They turn this ramshackle house into a proper house with windows and doors. They join together in a group and keep pestering the city authorities to supply electricity and running water to the shanty towns. Once they have been there long enough the authorities are sometimes willing to agree and then conditions improve greatly. But if people don't find work, little is likely to change.

e The candidate is back to writing quality answers. Options facing people who move into shanty towns are well understood. 2 marks.

(d) There are both bad and good consequences of the rapid growth of big cities in LEDCs, with more bad than good.

Pollution is one of the big problems. There is air pollution from the amount of traffic. Most buses, trucks and cars are too old to have catalytic converters and they belch out fumes, which are especially bad in Mexico City, because it is surrounded by mountains and the fumes remain trapped by the high ground. This leads to many illnesses and breathing problems among its inhabitants. Living in Mexico City has been described as 'like living in a gas chamber'. There is also water pollution. Factories take no notice of regulations, even where regulations exist. Waste that hasn't been treated runs straight into rivers which are like open sewers.

Traffic congestion is getting worse as more and more people can afford cars. As the city size increases with shanty towns around the edge, more buses are needed. The rush hour traffic in São Paulo is so bad that rich businessmen use helicopters to get from where they live to their offices in the CBD. For poor people there are not enough jobs and many people do not have a proper job. They have informal jobs with very low incomes, such as selling chewing gum on the streets in Rio.

Although city growth is mainly bad for the people living there, the good thing is that cities have offices and factories which employ people and make money for the country.

In São Paulo big companies like VW and Sony have factories, helping Brazil to become a more economically developed country, although it is not yet an MEDC.

 This is a top-level answer, which would earn all 7 marks for several reasons:
- The candidate realises that there are both good and bad consequences of big city growth. You should always look for both good and bad in questions which ask for consequences or impacts. Most candidates answering this question will only think about bad consequences, because there are more of them and they are better known.
- A good number of different consequences are referred to, including air pollution, water pollution, traffic congestion and unemployment, as well as industrial growth leading to economic development.
- Information that is precise is used — for instance, the candidate does not just refer to pollution, but states the types. Named examples of big cities are also used in the answer; this is good practice, followed by too few GCSE candidates.

Summary

What mark would the answer be given?
Almost full marks — 19 out of 20. This puts it towards the top of the A* band.

How could the answer have been improved?
It is a great achievement for any candidate to give six out of seven full-mark answers.

Answer to Question 8: candidate B

(a) (i) The total population has increased every 20 years.

 0 marks. Population growth is given in the question; the candidate is doing nothing more than repeating the question. Two population totals need to be quoted in order to describe the increase.

(ii) The highest number living in shanty towns was in 2000. Every time the population went up, so did the number living in shanty towns.

 The candidate seems unwilling to quote values from the graph, which is a big mistake. 1 mark.

(b) 1. Push factors from the countryside. Farm work is badly paid and there are better jobs in the cities. There are more hospitals and schools, so people migrate to the cities where these are found.
2. Pull factors of the cities. There are more jobs and more different types of work in the cities, e.g. jobs in factories and shops that don't exist in the countryside.

 In effect only one reason is identified — push and pull factors are both part of rural to urban migration. This reason is elaborated upon and so would earn 2 marks. The answer is not helped by the candidate mixing up push and pull

factors in the description of push factors. Industrial growth in the city could have been accepted as a second reason, if the candidate had stated it clearly.

(c) (i) What I see is a shack made out of wood, cloth and tin sheets. There are no windows in it and it does not seem even to have a proper door. It is what newcomers build for themselves by scavenging around when they reach the city without any money.

e This answer is based upon description from the sketch, which is good. An answer is less likely to go wrong when based upon direct use of the resource provided. 2 marks.

(ii) In a shanty town. In most big cities like Lima, shanty towns are located on the edges, where most empty land is found. People building shacks don't own the land. They are squatters on wasteland that no one wants to use, sometimes because it is too steep.

e A location is stated — on the edges. The rest of the answer helps to explain this location. 2 marks.

(iii) It might be bulldozed down and removed by the government. When this happens people often start building on the same land again once the bulldozers have gone. There is a game of cat and mouse between the shanty town dwellers and the government. The new houses that are then built may be no better than the old ones.

e The candidate gives an interesting answer and describes events that happen regularly in cities in LEDCs. It may not be the answer that was expected when the question was set, but it would still be considered on its merits by an examiner. The question asks for 'changes to the house'. In effect this answer skirts around the question, although it has some merit. 1 mark.

(d) Terrible problems are caused by the rapid growth of big cities in LEDCs:

- Massive pollution of the air from all the cars and factories.
- Forests around the edges of cities are cleared as newcomers search for land to build new shanty towns on.
- In Brazil the Amazon Basin is losing its forests at record rates as cities grow and extend into the forests.
- Wildlife habitats are being lost and species becoming extinct as the land is used for new houses and factories.
- New shanty towns have no sewers which is killing wildlife that remains after the forests have been cleared.
- Traffic congestion from thousands of buses and cars.

Some cities are trying to deal with these problems. One example is in Rio in Brazil where shanty towns are being cleared and people rehoused in new settlements outside the city. In Mexico City people are only allowed to drive into the city on certain days of the week according to the numbers on their number plates to reduce pollution and congestion.

 This answer has several weaknesses:

- Using bullet points is not an effective method of answering longer GCSE questions.
- The candidate is unable to resist including comments on housing, which are disallowed by the question.
- Some information is not accurate; the forests of the Amazon Basin are over 1,000 kilometres away from the growing cities of São Paulo and Rio, so their growth is not clearing the forests.
- The last part of the answer below the bullet points has nothing to do with consequences and is irrelevant.

This remains a low-level answer and much worse than the one given by candidate A, although there are enough references to consequences (air pollution, traffic congestion and loss of wildlife habitats around the cities) for it to earn 3 marks.

 Summary

What mark would the answer be given?
11 out of 20. A marginal grade-C performance, although some answers suggested the candidate was capable of achieving better.

How could the answer have been improved?
By quoting values from graphs and by not using bullet points in long answers.

Exam tips

1 Always quote and use values when describing from graphs.

2 Never use bullet points when answering longer questions.

3 When asked for consequences, look for both good and bad, even though one is likely to be more important than the other.

Urban zones in the UK

(a) Name **three** characteristics of the CBD of cities in the UK.　　(3 marks)

(b) Study the two sketches (Figures 1 and 2) of housing areas from the inner-city zone of UK cities.

Figure 1

Figure 2

(i) Name the main type of housing shown on each sketch.　　(2 marks)

(ii) Explain why these two types of housing are commonly located in inner-city areas of UK cities.　　(3 marks)

(c) During the last 20 years some inner-city areas in the UK have been redeveloped and improved. The London Docklands and Albert Dock in Liverpool are just two of many examples.

Choose **one** inner-city area you have studied that has been redeveloped.
(i) Describe how it has changed over the last 20 years. (5 marks)
(ii) Name a problem caused by these changes and explain why it has occurred. (2 marks)

Total: 15 marks

Answer to Question 9: candidate A

(a) 1. Shops and offices dominate the CBD. This makes it the main place of work during the day with traffic congestion in rush hour.
2. Buildings are higher and big cities have skyscrapers like Canary Wharf in London.
3. Old buildings, such as cathedrals and churches, are in the centre as well.

𝑒 The command word 'name' means that answers do not need to be as long as these. However, when you know and understand, the best policy is to give more than the basic answer, in case one answer is not considered acceptable. For example, this candidate has named two characteristics under number 1, giving one spare that could have been credited if needed, although in this case it wasn't. 3 marks.

(b) (i) Figure 1: terraced housing. Figure 2: blocks of flats.

𝑒 Correct answers. 2 marks.

(ii) Terraced houses are the oldest type of houses found in British cities. They were built to house workers at the time when industries were located in city centres. Then everyone lived in terraced houses (as in Coronation Street in Manchester). Some terraced houses were pulled down in the 1960s and replaced by the blocks of flats shown in Figure 2, so that as many of the people who lived there as possible could be rehoused in the same area.

𝑒 This is a well-focused answer which contains explanation for both types of housing. 3 marks.

(c) (i) My chosen area is Manchester Docklands near to where I live, which has changed greatly and is still changing. Derelict old warehouses in Castlefields have been modernised and made to look good again. One has been changed into a hotel called the Victoria and Albert, although most have become flats and apartments. The canal basin has been cleaned out and there are paths, cafés and restaurants along its sides. It is now a visitor attraction and is near another attraction that has been there for longer, the Granada TV studios tour that includes the old set for *Coronation Street*.

Visitors can now go on a boat ride on the Irwell to Salford Quays. They use a boat like that on canal tours in Amsterdam. The River Irwell used to be known as the local sewer. Now high-priced apartments and prestige offices are located along its banks.

 The message is — if you know a local example, use it. Case studies based on a candidate's home area almost invariably include more names and specific information about the place, as this one does, which allow marks to be clocked up quickly. There is more than enough content in this answer for all 5 marks. Local colour makes the answer more interesting and this candidate is obviously a *Coronation Street* fan! However, interesting answers are only given the marks when the relevant content is included.

(ii) I don't think there are any problems now. This area is much better than it used to be, because it looks better and is attracting more people back to live in the centre of Manchester, which is known as gentrification. It is also good for Manchester's image.

 This is where examining becomes heartless — the question asked for a problem and the candidate is unable to name one. Therefore, the answer would receive 0 marks, despite the candidate's knowledge of gentrification. In most inner-city redevelopments, advantages greatly outweigh disadvantages, but it is unlikely that no problems exist. For example, expensive apartments can only be afforded by the wealthy and there are no new homes for former inner-city residents, who are moved to other parts of the city. This is one of the general problems associated with gentrification that this candidate could have used.

Summary

What mark would the answer be given?
13 out of 15. Clear grade-A standard.

How could it have been improved?
It couldn't, except for the lack of a problem in the final part.

Answer to Question 9: candidate B

(a) 1. Many shops and offices.
   2. Bus and train stations.
   3. The busiest part of the city.

 In the first two answers the candidate names CBD characteristics, but the third answer is too vague for a mark. Out of working hours the CBD may be the least busy area. Giving brief answers obeys the command word 'name', but there is nothing spare if one of the answers is not accepted. 2 marks.

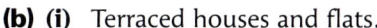

(b) (i) Terraced houses and flats.

e This would earn 2 marks, but it would have been safer to have stated the figure number for each one.

(ii) These are old types of houses, built for factory workers. They were built so that people could live close to their places of work in the city centre. Inner-city areas are the oldest part of the city. There is a problem now because more people own cars and there is nowhere to park them other than on the street. A lot of these old houses are now being pulled down and people are moving to the suburbs.

e There are two problems with this answer:
- There is no separate mention of terraced houses and flats. The comments apply to terraced houses only.
- The second half of the answer is about problems and is irrelevant.

The first three sentences, which explain the presence of terraced houses, would gain 2 marks.

(c) (i) London Docklands. It used to be all docks, but when the docks were closed and no longer needed, warehouses were made into flats. Some of these have good views over the River Thames and are popular with people. This area now has a high density as people prefer to live here next to the river than in other parts of London. It also attracts a lot of tourist visitors because it is new and different. The Dome was built here on wasteland next to the Thames but that is now empty.

e After the first sentence, it soon becomes clear that the candidate is struggling to find real information about the London Docklands. This is not a case study answer based on strong local knowledge, as the one from candidate A was. This is a low-level answer which would only get 1 mark, for the point about warehouses being converted into flats. The London Docklands is a popular case study choice, used by many GCSE candidates in exam answers. Many of their answers include mention of office developments in and around Canary Wharf, newspaper offices and new forms of transport such as the Docklands Light Railway and London City Airport.

(ii) High cost of houses in London. Poor people cannot afford them.

e A valid problem is named, which would gain 1 mark. Otherwise the answer given is a general one, not directly related to the London Docklands.

Summary

What mark would the answer be given?
8 out of 15. This is on the borderline between grades D and C. It is not a secure grade C as most of the marks were gained on the shorter, easier questions.

How could the answer have been improved?
By greater case-study knowledge for an area of inner-city redevelopment.

 **Exam tips**

1 Good case-study knowledge is essential for grade-A answers and helps to make grade-C answers more secure. Less able candidates tend to pick up a high proportion of their marks from the short-answer questions, leaving case-study answers to discriminate between candidates above and below grade-C level.

2 If there is a case study that fits the question from your home area, use it. Your knowledge of it is likely to be greater and more certain.

Farming in the UK

(a) (i) State the difference between arable and pastoral farming. (2 marks)

(ii) What is meant by intensive farming? (2 marks)

(b) Study Figures 1 and 2 which show the same farm in East Anglia in 1950 and 2000.

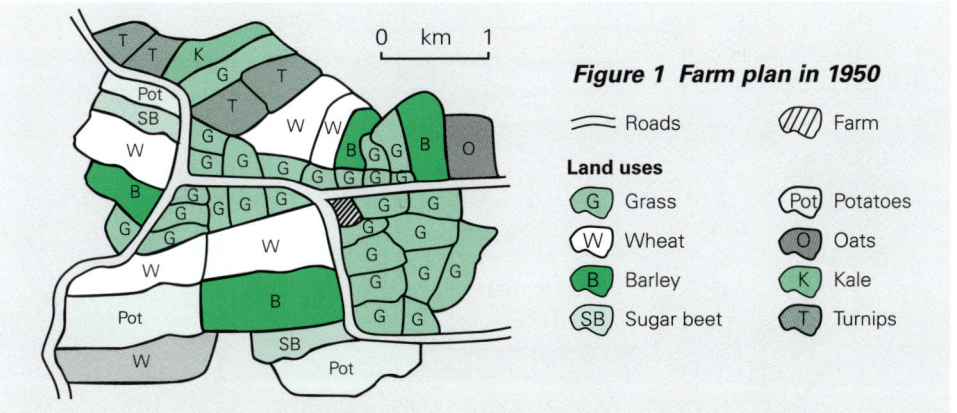

Figure 1 Farm plan in 1950

≋ Roads   ▨ Farm

Land uses

G Grass   Pot Potatoes

W Wheat   O Oats

B Barley   K Kale

SB Sugar beet   T Turnips

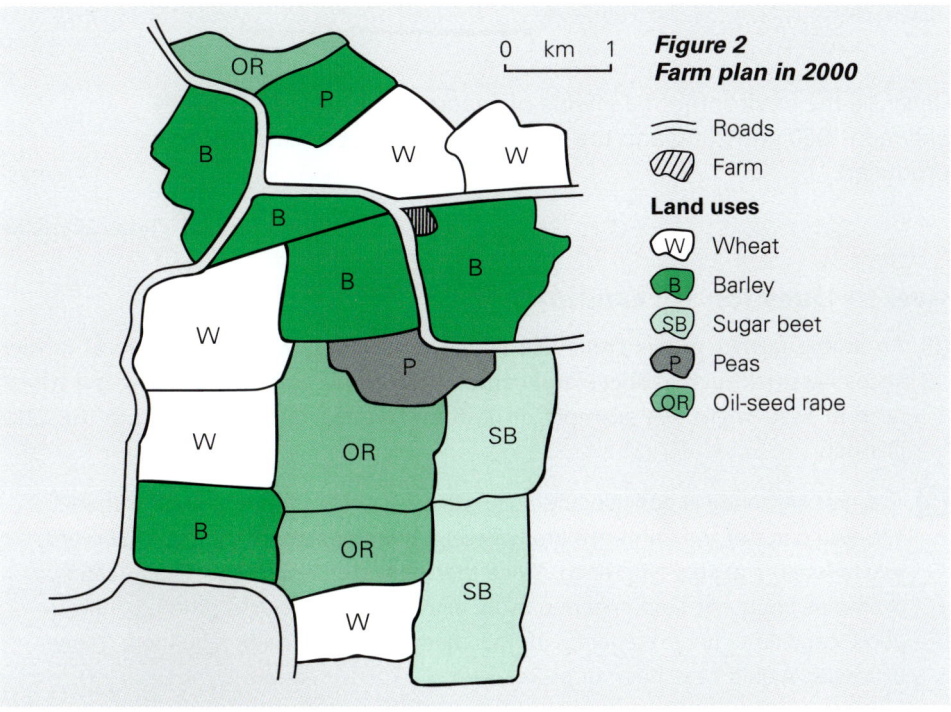

**Figure 2
Farm plan in 2000**

≋ Roads   ▨ Farm

Land uses

W Wheat

B Barley

SB Sugar beet

P Peas

OR Oil-seed rape

(i) Name the land use which covered the largest area of land in 1950. (1 mark)

(ii) In 1950 the farm was a mixed farm. Describe the evidence from Figure 1 which supports this. (2 marks)

(iii) By 2000 the farm was an arable farm. Describe the evidence from Figure 2 which supports this. (3 marks)

(c) Figure 2 shows that farm size and field sizes have increased greatly from 1950 to 2000. Suggest reasons why farm and field sizes have increased in many parts of the UK. (3 marks)

(d)

Describe the environmental effects of changes in farming in the UK between 1950 and 2000 and the conflicts of interest which they have produced. (7 marks)

Total: 20 marks

Answer to Question 10: candidate A

(a) (i) An arable farmer grows crops like wheat and barley, whereas a pastoral farmer keeps livestock such as sheep and cattle. An example of arable farming is a wheat farm in East Anglia. An example of pastoral farming is a sheep farm in the Lake District.

🄮 The first sentence is concise and accurate, and it would earn the full 2 marks. The two parts of the sentence are linked by 'whereas', which is the best word to use when stating a difference. Since there are only 2 marks for this question, examples could not be demanded by an examiner, but including them confirms good candidate understanding. If this had been a 4-mark question, these examples would have been of great value.

(ii) This is when the farmer gets a high output from farming his land. High outputs are obtained by working hard and spending money on fertilisers and good seeds. Market gardening is an example of intensive farming.

e 2 marks. The candidate realises that intensive farming is all about high inputs in order to gain high outputs. The candidate answers in a similar way to that used successfully in the previous question and states another suitable example.

(b) (i) Grass.

e 1 mark.

(ii) Mixed farming is when a farmer grows crops and keeps animals. The farmer in 1950 was growing crops such as wheat, barley, sugar beet and potatoes. He must have been keeping animals because of the amount of grass grown, although no animals are named in the key.

e 2 marks. The candidate makes clear his/her understanding of what is meant by mixed farming and uses the information in Figure 1 to make pertinent points. The candidate could also have made the point that crops such as oats and turnips, as well as grass, were grown in 1950 for animal feed.

(iii) An arable farmer grows crops only. Only crops were being grown on this farm in 2000. Wheat, barley, sugar beet, oil-seed rape and peas were the five main crops being grown. The grass fields had all gone and the fields were larger.

e The candidate approaches the answer in the same well-organised way as in the previous questions, but this time does not manage to go far enough to gain all 3 marks. The question intended that candidates should go beyond the obvious points to secure the third mark. This candidate notes that fields are larger, but fails to take the answer further by describing how this helps arable farmers, for example by allowing more and larger machines to be used. 2 marks.

(c) Farms have joined together, so those that are left are bigger. This has happened as farmers have gone out of business and migrated to the towns for better standards of living. Farmers have taken out the hedgerows which is why field sizes have increased. Some people have called this 'prairieisation' and many conservationists are worried about this because a lot of wildlife habitats have been lost in the UK.

e There is some truth in the first sentence about why farm sizes have increased, which would earn 1 mark. The candidate could have elaborated by giving reasons why some people have left farming, such as low incomes and problems like BSE. After this, the candidate is correct in saying that field sizes have increased because hedgerows have been taken out, but is not answering the question. Reasons why the hedgerows have been removed should have been suggested, such as to allow the use of big machines and increase the area of land that could be farmed profitably. The point about loss of wildlife habitats is irrelevant in this part of the question; the candidate is starting to give an answer more useful to the next part, (d). This is a reminder that you should always look ahead to the next part of a question — two parts of the same question are not going to have the same answer. For a variety of reasons, therefore, this answer would only be given 1 mark.

(d) Agribusiness is the name given to farming in many parts of East Anglia. The farm owner's comments are likely to come from someone working in agribusiness. Agribusiness is when someone is running the farm to make a good profit from it. This is an example of intensive farming because a lot of money is spent on machinery, good seeds, fertiliser and pesticides and farm owners must obtain high outputs and good yields to make a profit.

Prairieisation means that the farmland now looks like that on the Prairies of Canada with big fields and no hedgerows between them. The appearance of the countryside in places like East Anglia has changed. Conservationists do not like what has happened. Many animals and plants that lived in hedgerows have now lost their habitats. There is nothing to stop strong winds sweeping across the land and blowing away the top soil. This causes soil erosion.

Another environmental effect of the changes not mentioned by the conservationist is pollution of rivers and streams from chemicals used on the farms. One example is nitrates that are washed out of the soil by rain and end up in rivers, where they are concentrated and kill fish and plant life in the water. They also pollute our water supply.

The conflict of interest is between farmers, who want to make big profits from farming, and conservationists, who want to protect the environment. As field and farm sizes have increased in East Anglia, the conflict has got worse. One of the new policies of the CAP is to pay farmers for looking after the countryside. This is called 'stewardship' of the countryside and this might reduce the conflict in future.

e This candidate continues with his/her planned approach to answering.
- The first paragraph refers to the farm owner and what he said.
- The second paragraph refers to changes in the countryside and what the environmentalist said.
- In the third paragraph details of a separate environmental effect are given, which extends the answer.
- The comment at the end refers to the 'conflict of interest'.

This is effective answering technique for longer questions; it ensures that all aspects of the question are covered and that nothing is missed out. The other good feature of this answer is the candidate's knowledge and understanding of relatively new terms, such as 'agribusiness', 'prairieisation' and 'stewardship' of the countryside. This answer has the depth and breadth of coverage to earn all 7 marks.

Summary

What mark would the answer be given?
17 out of 20. A strong grade-A answer, greatly helped by the quality of the answer to the final part which was worth more than one third of the total marks.

How could the answer have been improved?
The only 'sticky patches' were in answers in the middle. Taking more notice of the number of marks in (b)(iii), and of the question theme in (c), might have improved answer quality.

Answer to Question 10: candidate B

(a) (i) Arable farming is growing crops. Pastoral farming is keeping animals as well.

> *e* This answer has been spoilt by the candidate adding 'as well' at the end of the second sentence as this could be the definition for mixed farming. Does the candidate clearly understand the difference between the two types of farming? It is not clear from his/her answer. 1 mark.

(ii) Rice farming in monsoon Asia is an example of intensive farming.

> *e* This answer is difficult to mark. The example is correct, and often an examiner will give credit for a candidate using a correct example in a definition; however, this is usually after a definition has been attempted. Since this candidate attempts no definition of intensive farming, the likely mark would be 0.

(b) (i) Fields, with grass and crops like wheat and barley in them.

> *e* 0 marks. There can only be one land use which covers the *largest* area of land; therefore, grass is the only acceptable answer. The inclusion of 'fields' suggests that the candidate might not understand the meaning of 'land use', which is what the land is used for.

(ii) The farmer was growing grass and crops like wheat and barley in 1950. Animals that could graze on the grass would be cattle and sheep, so the farmer would have made some money from selling milk and meat as well as crops.

> *e* The candidate mentions both crop growing and keeping animals, the essential parts of a mixed farm. 2 marks.

(iii) Five different crops are being grown. There are cereals, mainly wheat and barley. Wheat is grown in five fields and barley is grown in four fields. Oil-seed rape grows with a bright yellow flower and its oil is used for making margarine. Sugar beet is being grown in two large fields south of the farm. One vegetable is grown, which is peas. The fields in which these crops are growing are much bigger than those in 1950. This is because tractors and combines are much larger than they used to be.

The candidate is making hard work of answering the question. Using six sentences to convey the message that only crops are being grown makes the answer look fuller than it really is. However, on the positive side the candidate is trying to use evidence from Figure 2, even if he/she doesn't know the best way of doing this for the question. There is also some merit in the point made in the final sentence about field size and the use of machinery. The main weakness of the answer is that the candidate does not indicate how the evidence quoted shows that this is an arable farm. 2 marks.

(c) Farmers have pulled out hedges to make the fields larger. They use large tractors and combines that need more room to work in and to turn. It would be impossible to use the big machines in the small fields shown in Figure 1. Farm work is done more quickly and easily when fields are large.

 Comments made by the candidate about field sizes are valid, but no mention is made of increased farm size. Did the candidate not notice farm size when reading the question, or was the answer not known? This would result in the loss of 1 mark, which would leave this as a 2-mark answer.

(d) People say that modern methods of farming do a lot of damage to the environment. Heavy machines press on the ground and compact it, so that when it rains water can no longer soak into the ground and it washes off the surface, causing soil erosion. Another way in which soil erosion is caused is by strong winds after the hedgerows are removed. Hedges act as windbreaks and protect the soil on both sides. People who like to see wild birds and wild animals, complain that habitats have been lost as hedges have been pulled down to make larger fields, and as marshy areas have been drained for cultivation. They also say that most farmers use too many chemicals like insecticides which poison many small creatures and reduce the wildlife.

 The candidate describes several different effects from changes in farming, but this is the only part of the question attempted. There is no mention of conflicts of interest between farmers and other groups of people which formed the second part of the question. The quotes from the farm owner and conservationist are included to help with the answer and referring to them could have helped the candidate produce a broader answer. This answer would fall in the middle of the mark range and be worth 4 or 5 marks. It is an unbalanced answer, with one part of the question answered well and the other ignored.

Summary

What mark would the answer be given?
11 or 12 out of 20. Grade C standard. In many ways this is a typical grade-C answer because it is good in parts, but not throughout.

How could the answer have been improved?

By more careful reading of the questions and by giving more considered answers which covered all aspects of the questions. The mark is probably below what would be expected for this candidate's knowledge and understanding of UK farming.

 Exam tips

1 When there are two aspects to a question, make sure that you answer both.

2 Use all the information given in the question; it is included to help and to trigger the best response.

11

The green revolution

(a) State what is meant by the green revolution. (3 marks)

(b) Study Figure 1 which gives information about rice output in Asia from 1960 to 2000, relative to the output in 1960.

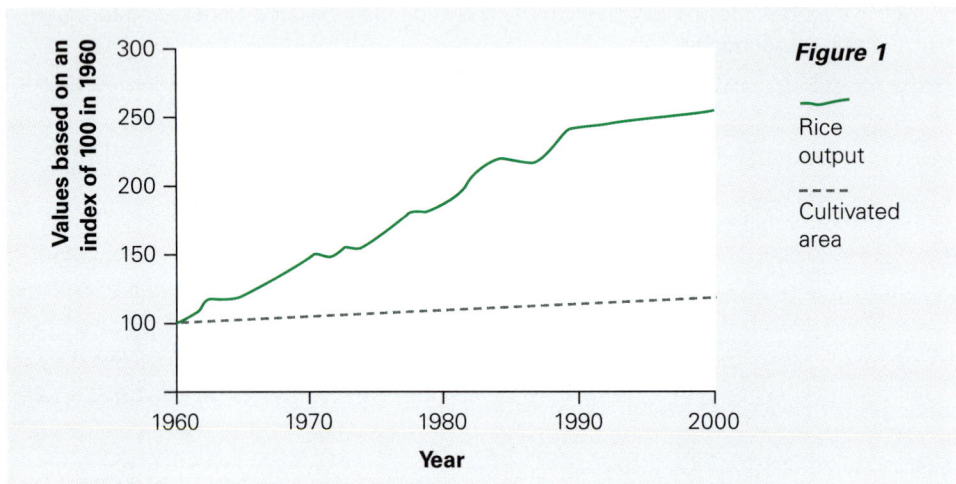

Figure 1

— Rice output

- - - - Cultivated area

(i) Describe what has happened to rice output in Asia since 1960. (2 marks)

(ii) How does the information in Figure 1 suggest that the green revolution has been widely introduced in Asia? (2 marks)

(c) Explain why not all farmers in LEDCs have benefited from the green revolution. (5 marks)

Total: 12 marks

Answer to Question 11: candidate A

(a) (i) The most important feature of the green revolution is the use of new seeds that give higher yields than the old seeds. IR8 is an example of a new variety of rice seed that is now widely used in India. It has been called a miracle seed because of the amount it produces compared with the old seeds.

e The candidate is correct in stating that new seeds are the most important feature of the green revolution. The named example of a type of seed is also correct. These points would earn the candidate 2 marks, but another characteristic of the green revolution is needed because this is a 3-mark question.

Mention of the greater use of fertilisers, or machinery, or irrigation, all examples of the modernisation of farming associated with the green revolution, would have earned the third mark.

(b) (i) It has gone up. In fact, it has more than doubled between 1960 and 1990, although the line has levelled out and it has not gone up as quickly from 1990.

e Everything is included that you would expect in an answer to this type of question:
- The general trend is stated — 'it has gone up'.
- Reference is made to values — 'it has more than doubled between 1960 and 1990'.
- Differences within the general trend are noted — 'it has not gone up as quickly from 1990'.

This is the correct way to approach the description of a graph and the answer would receive the full 2 marks, with something to spare.

(ii) I think this is shown by the differences between what has happened to the two lines. I have already described how the rice output has gone up, but the line showing cultivated land has stayed the same, or almost the same. There may have been a little increase, but nothing like as much as the output. This suggests that farmers must be using new seeds to get a higher output of rice from the same amount of land.

e In the first sentence the candidate shows that he/she understands the question and knows how to answer it. The rest of the answer confirms this understanding. 2 marks.

(c) Not all farmers can benefit from the green revolution because many cannot afford to buy the seeds to start them off. Farmers could keep their old seeds from one year and use them the next year. Now they have to buy new seeds and many don't have the money to do this. Some farmers were given the new seeds by the government and found that they got lower yields than from their old seeds. This was because they continued to farm using the same methods that they had always used and didn't learn modern methods of farming. They were too poor to do this.

e The candidate relies heavily upon one reason why not all farmers have benefited from the green revolution — farmer poverty. Some elaboration upon this reason is given, but little information of a precise nature is stated and the candidate is not stating much that is new or different in the second half of the answer. The candidate could have developed the answer by mentioning that many farmers borrowed money for seeds, fertiliser and machinery and ended up with large debts that they are unable to pay off. In a good answer candidates might have been expected to refer to the essential need for irrigation water, pesticides and fertilisers to ensure successful growth of high-yielding varieties.

In essence this question is asking for the disadvantages of the green revolution and the candidate does not recognise this. Overall, the candidate's answer lacks the content required for a 5-mark question. 2 marks.

Summary

What mark would the answer be given?
8 out of 12. Top grade B, 1 mark short of a grade A.

How could the answer have been improved?
By increased content in the two answers to those parts of the question with most marks attached to them.

Answer to Question 11: candidate B

(a) New seeds that give better yields.

ℓ This is a basic 1-mark answer. The candidate only makes one point in answer to a question worth 3 marks, so there is no way 2 of the 3 marks could be claimed.

(b) (i) It has increased by a large amount. In 1960 the value was 100 and it has gone up and up to 2,000.

ℓ Another basic 1-mark answer. To claim the second mark the candidate needed to use at least two values from the graph to show that output had increased.

(ii) To have increased the output, farmers in Asia must have used the new seeds. The output line has gone up while the dotted line showing the cultivated area is constant.

ℓ The candidate refers to both lines, but the difference between the two lines is stated without being commented on in relation to the question. The candidate hints at knowing the answer without managing to show that he/she does. This restricts the value of the answer to 1 mark.

(c) Rich farmers have benefited most in India. They could afford to buy the new seeds and the fertilisers they needed to make sure that they grew well. Some farmers have had bumper crops of rice which has made them even richer. Poor farmers borrowed money to buy seeds and fertilisers, hoping to grow so much more rice that they would have some spare to sell in local markets and make enough money to pay back their loans. In Bangladesh many farmers are even poorer than those in India because of the many floods. Fewer of them have been able to afford to use the new seeds and there is less of a green revolution in Bangladesh than in India.

ℓ This is definitely this candidate's best answer; for the first time an answer of decent length is given. Comments about differences between rich and poor

farmers in India are useful. Mention of differences between poor farmers in India and Bangladesh is also relevant. In the majority of African countries even fewer farmers have access to new seeds than in Asia. This answer would be worth 4 marks because it contains a greater amount of precise information than the one from candidate A.

Summary

What mark would the answer be given?
7 out of 12. Just a grade C, saved by the answer to the final part. Answers to the short questions fall short of this grade.

How could the answer have been improved?
By giving longer answers to the short questions and taking more notice of the number of marks for each question.

Exam tips

1 Answering well the longer, more knowledge-based questions is a great boost to the grade achieved. The opposite also applies — answer them badly and grade hopes suffer.

2 Length of answers when giving definitions and describing from graphs must reflect the stated number of marks for the question.

Employment and manufacturing industry

(a)

Farmer

Shopkeeper

Miner

Hotel worker

Steel maker

Worker in a car factory

From the above list, choose the **two** jobs from each sector of employment.
(i)   Primary
(ii) Secondary
(iii) Tertiary   (3 marks)

(b) Study the **two** pie graphs below. They show average percentages for employment sectors in LEDCs for 1960 and 1995.

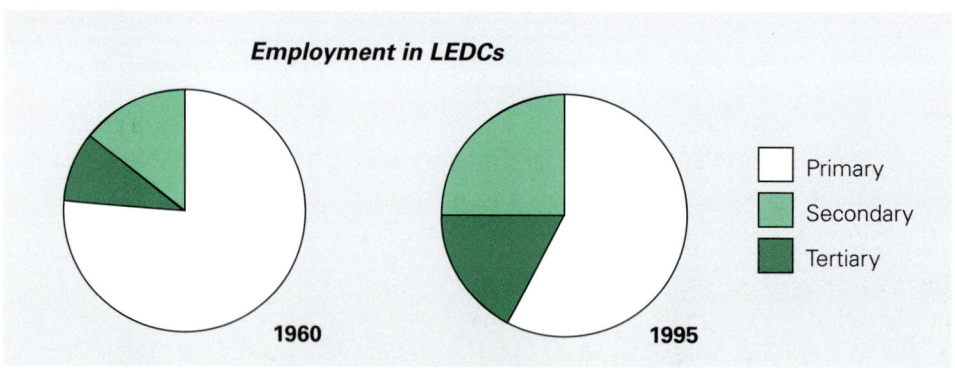

Employment in LEDCs

1960   1995

☐ Primary
☐ Secondary
☐ Tertiary

(i)   Describe the changes shown from 1960 to 1995.   (3 marks)
(ii) How do they show that LEDCs have become more economically
      developed?   (2 marks)

(c) Choose **either one** named area in the UK **or one** LEDC with many manufacturing industries. Explain why these industries have grown in the area or country chosen.   (7 marks)

Total: 15 marks

Answer to Question 12: candidate A

(a) (i) Farmer and miner.

(ii) Steel maker and worker in a car factory.

(iii) Hotel worker and shopkeeper.

e All the answers are correct. 3 marks.

(b) (i) Primary has gone down from 75% to about 60%. Secondary and tertiary have gone up. The tertiary has gone up to just over 25% in 1995 from about 16% in 1960. More people are living in towns and fewer are staying on the land to farm.

e There are two strong points about this answer:
- It includes references to changes in all three sectors of employment.
- Values have been used to support the changes described, in this case percentages. Fractions, such as three quarters and one quarter, would have been equally acceptable.

The answer would earn all 3 marks. However, there is no credit for what is stated in the final sentence, because the candidate is no longer describing the changes shown in the pie graph, as required by the question. Irrelevant information like this is normally ignored by an examiner.

(ii) When a country becomes developed, it has a lot of industry. People move to cities (this is called rural–urban migration) to work in factories, shops and offices, where wages are higher. This leaves fewer farmers (primary sector) and the economy of the country depends less on farming as well.

e This answer contains relevant information. It shows that the candidate understands that work in the primary sector, such as farming, is replaced by work in factories, shops and offices in the secondary and tertiary sectors as a country becomes more economically developed. 2 marks.

(c) There are many light and high-tech industries along the sides of the M4 between London and Bristol. One area is on the Slough industrial estate where Mars bars are made and there are other food-processing factories, such as those making soft drinks. It is right next to the M4 for easy access to markets. High speed trains run along the line between Bristol and London. There are also high-tech industries in this area, which need skilled workers, who come from the many universities in the area, such as Bristol and Reading. Road and rail links to London are excellent, where the largest market in the UK is found. They also go to Heathrow airport, the UK's largest airport, with good links to countries and markets overseas. All these factors explain why industries have grown in this area.

e There are several strong points about this answer:
- The candidate starts in the best possible way for a case study question by naming an appropriate area.

- All the information given clearly relates to the named area.
- Some precise details are included on transport and the source of skilled workers.
- The significance of some points is suggested, such as of Heathrow airport.

The use of specific information for a named area places this answer in the top level. It would gain 6 marks. A little extra detail would be needed to claim all 7 marks. This could have been supplied by giving more detail about factors already described, such as road links from the M4 and M25 to other parts of the UK, or the size and wealth of the market in London and southeast England. Another way would have been to refer to an additional factor, such as nearness of pleasant countryside (e.g. the Chilterns and chalk Downs), which could help in attracting skilled workers to the area.

Summary

What mark would the answer be given?
14 out of 15. This is an example of an A* answer.

How could the answer have been improved?
Only by including more case study information in part (c).

Other examples of answers to Question 12(c)

Candidate B (for an industrial area in the UK)

Industries need lots of workers. Good transport such as roads and railways are also needed. High-tech industries are growing in the UK because more and more people are using computers. There are plenty of these industries next to the M4 near London because of good transport and big markets.

 This is a poor answer because:
- The area chosen is named only at the end.
- No precise information for this area is given.
- Vague terms such as 'good transport' are used that are of little value.
- The answer is much too short for a 7-mark question.

It would only receive 1 of the 7 possible marks. The effect of a poor case study answer like this would have been to reduce the total mark for the answer from 14 out of 15 to 9 marks (assuming that other answers were of similar quality to candidate A's). This would change the value of the answer from grade A to grade C, which highlights the importance of using case study information that is specific in longer GCSE answers.

Candidate C (for an LEDC)

An LEDC with many manufacturing industries is South Korea. Its main industries are shipbuilding, steel, cars, electronics and electrical goods. It builds more ships than any other country in the world. Most industries are located in and around Seoul, its capital city. Cheap labour is the most important reason why industries have grown here; the average wage rate is around five dollars an hour, about one third of that in the US. People also work hard and for long hours. Despite the low wages, the workforce is quite skilled, as people in South Korea are better educated than in most LEDCs. The government puts tariffs on the import of manufactured goods from overseas, which have encouraged the growth of big companies like Samsung and Daewoo. Many of the goods made by these big companies are exported and they are close to large markets in Japan and China. Markets for manufactured goods are growing fast in countries around the sides of the Pacific Ocean. Goods can be exported easily from the port of Pusan to all parts of the world. Most industrial goods are now put in containers, which is a cheap and safe way of sending goods by sea.

e This answer would be given the full 7 marks because:

- Specific information about South Korea is used throughout the answer.
- Several different factors are referred to, such as labour, government and markets.
- The significance of the factors is explained.
- There is enough content for a 7-mark answer.
- It is organised in a clear and effective manner and is well written.

Exam tips

1 Case studies are important. A weak case-study answer can reduce the worth of the final answer by two or more grades.

2 Case-study answers do not need to be long, but to be effective they must contain specific information.

Question *Economic activity*

Business parks in the UK

(a) Study the photograph below which shows a business park built on the edge of an English city.

(i) State **three** characteristics of the business park shown in the photograph. (3 marks)

(ii) Choose **one** of the characteristics you have stated in part (i) and explain how it suggests that this business park is located out of town. (2 marks)

(b) Many environmental groups and the UK government favour the building of new business parks on brownfield sites nearer the city centre instead of greenfield sites outside the city.

(i) What is a brownfield site? (1 mark)

(ii) Explain some of the arguments against building new business parks on greenfield sites outside the city. (6 marks)

Total: 12 marks

Answer to Question 13: candidate A

(a) (i) 1. A big, modern building.

2. Car parking space and plenty more space around it.

3. Nike has an office here, which is a transnational company.

e The candidate rightly concentrates on stating characteristics that can be seen in the photograph. This candidate has in fact stated five valid characteristics for a 3-mark question — both 'big' and 'modern' under statement number one, and 'car parking space' and 'plenty more space' under statement two. This is good practice, because it maximises the chances of claiming all the marks should one or more of the characteristics not be accepted. 3 marks.

(ii) Number 2. Car parking places and all the other space don't exist in the city, where there is much more pressure on land and land costs more than on the edge. This is one of the main advantages for a business locating out of town.

e The candidate makes very clear which one of the characteristics has been chosen and gives another clear answer. 2 marks.

(b) (i) An area of wasteland. There is more wasteland in inner-city areas than anywhere else.

e A more precise definition of a brownfield site is 'land that was previously used and built on'. By mentioning waste and suggesting an inner-city location, the candidate has come close enough and so would be given 1 mark.

(ii) There are many advantages to building business parks on greenfield sites. Land is cheap and plentiful. It is easy for people to get to work, because it is close to motorways that mostly run around the edges of towns. More people now live in the suburbs and in rural areas so that they are closer to where they work than they were when all offices were located in the CBD. They have shorter and easier journeys to work by car when they work on a business park. Big trucks also find business parks easier to reach because they can avoid the traffic congestion in CBDs.

There are also some disadvantages. Trees are cut down and more countryside is lost. Environmentalists do not like this because more wildlife habitats are destroyed and some species of field birds are in danger of becoming extinct. Some people complain that the UK is becoming a concrete jungle as one development around the edges of cities joins up with another. It should be left as green-belt land.

e The first paragraph is irrelevant. The shorter second paragraph is better, because the candidate refers to some environmental disadvantages. The question this candidate is answering is 'Explain the advantages and disadvantages of...', instead of 'Explain the arguments against...'. Advantages are worth no marks; disadvantages can be given some marks, but not as many as arguments. This answer would be worth 2 or at best 3 marks.

Summary

What mark would the answer be given?
At least 8, probably 9, out of 12. It is on the borderline between grades A and B.

How could the answer have been improved?
By the candidate concentrating upon giving arguments against using greenfield sites, instead of changing the question into advantages (irrelevant) and disadvantages (not quite the same as arguments).

Answer to Question 13: candidate B

(a) (i) The business park is built on cheap land. Parking cars is easy and people do not have as far to travel from their homes. They are popular places for people to go to work. They are located near to motorways for fast access by car.

e Has the candidate looked at the photograph? Certainly he/she has paid little attention to the instruction to state characteristics shown in the photograph. The closest the candidate comes to stating anything that can be taken from the photograph is in the reference to 'parking cars'. The answer would probably not be given any marks, although another examiner might give it 1. It is a classic example of how *not* to answer a question based on observation from a photograph.

(ii) Land is cheap because there is less competition than in CBDs, where all the shops and offices are found.

e The characteristic stated is not taken from the photograph. Therefore this is not a valid answer to the question. 0 marks.

(b) (i) Fields with no crops growing in them near the city centre.

e 0 marks. The candidate does not know what the term 'brownfield site' means and has tried to work out from the question what it might mean without success.

(ii) Some of the arguments of environmentalists against using greenfield sites are as follows.
A Forest and vegetation removal. They say that this has reached levels that are too high already as more and more of the countryside is built on.
B Habitat loss for wildlife. Birds and animals have fewer places to live, reducing their numbers in some cases to the point of extinction.
C Loss of farmland and places for recreation. Beautiful countryside is lost where people spend leisure time at weekends.
D More pollution. People working on business parks outside towns often have no choice but to travel there by car, increasing air pollution and releasing more carbon dioxide by burning fossil fuels.

I think that these are good arguments against using greenfield sites because the amount of countryside cannot go on being reduced for ever. The government wants to reduce the output of greenhouse gases to meet its targets from the Kyoto agreement, which it is not going to do if more and more people travel to work by car to new business parks in the countryside. It also wants to stop cities merging together, and green belts around them are supposed to stop this.

 This answer has several strong points:

- The candidate concentrates on putting forward arguments.
- Mention is included of both groups named in the question (environmentalists and government).
- A number of different arguments are used.

Often lettering different points A–D is not the best way of answering longer questions, but here it seems to help by encouraging the candidate to make clear the different arguments. In addition, the points are explained and not just listed. 6 marks.

Summary

What mark would the answer be given?
6 out of 12. A top grade D, held back by the disastrous answers in part (a), for which the very good answer to the final part (b)(ii) cannot fully compensate.

How could the answer have been improved?
By using the correct technique for answering photograph-based questions.

 ## Exam tips

1 Careful observation is essential in photograph-based questions; without it, an effective answer is not possible.

2 Giving arguments for and against is not the same as giving advantages and disadvantages. Some of the same information can be used, but it needs to be presented differently.

Energy sources in the UK

(a) Study Figures 1 and 2 which show how electricity was generated in the UK in 1990 and 2000.

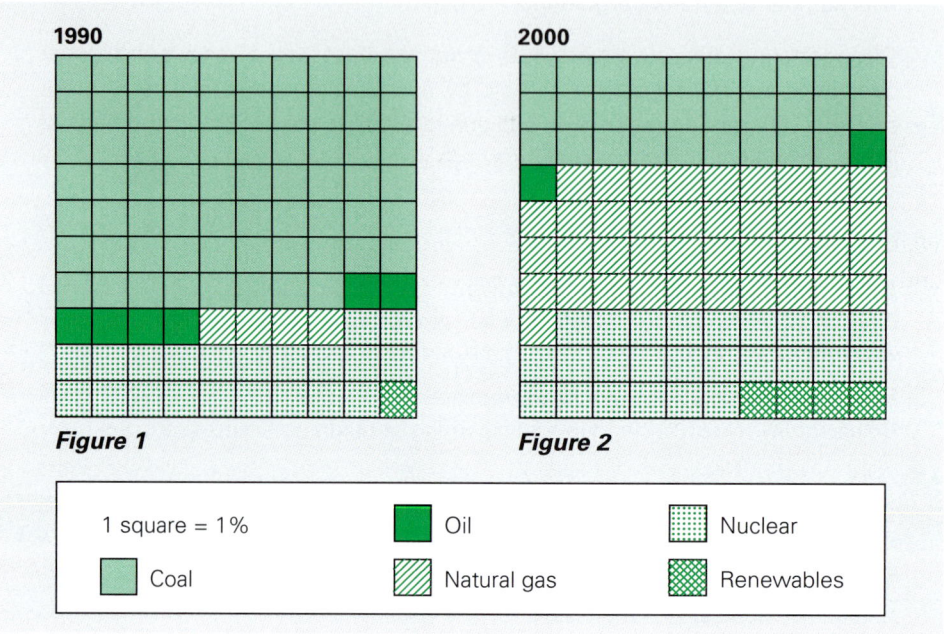

Figure 1   **Figure 2**

1 square = 1%   Oil   Nuclear
Coal   Natural gas   Renewables

   (i) State the percentage reduction in coal use from 1990 to 2000. (1 mark)
   (ii) By how many times did the use of natural gas increase from 1990 to 2000? (1 mark)
   (iii) List the energy sources in order of importance for making electricity in 2000. (1 mark)

(b) Explain why natural gas has become more important than coal for electricity generation in the UK. (4 marks)

(c) (i) Name and locate **one** example of a renewable source of energy used to generate electricity in the UK. (2 marks)
   (ii) The UK government is keen to increase the amount of energy generated from renewable sources, but in 2000 only a small percentage of the energy used was from renewable sources. Explain this statement. (6 marks)

Total: 15 marks

Answer to Question 14: candidate A

(a) (i) 39

e This would probably get 1 mark, although the mark is only secure when the unit is stated. 39% would have been better.

(ii) 10 times.

e 1 mark.

(iii) Coal first followed by nuclear, oil, natural gas and renewables.

e 0 marks. The candidate has given the order of importance for 1990, not 2000. This is the kind of careless mistake which it is easy to make in an examination. Checking every answer reduces the chances of it happening.

(b) Natural gas is clean and cheap. The UK has its own natural gas fields in the North Sea and Morecambe Bay. The change towards using more gas to generate electricity has been called the 'dash for gas'. Governments in the UK encouraged this after they had attended the Kyoto summit in 1995, because natural gas gives out only small amounts of carbon dioxide when it is burnt. This is helping to reduce the amount the UK contributes to global warming.

e This is a good answer to the natural gas part of the question. The candidate demonstrates a clear understanding of the 'dash for gas' and uses precise information supported by named references to places. However, the lack of any mention of coal is a major weakness, because coal is also named in the question. To earn full marks, the candidate also needed to demonstrate why coal is not as good: for example, why it is dirtier to burn and more expensive to obtain than natural gas. 3 marks.

(c) (i) Wind farm. There is one close to where I live right next to the coast of Cumbria on wasteland north of Workington.

e 2 marks. The example is correct and the location is stated clearly. This illustrates the advantage of using an example from your home area when this is possible.

(ii) Renewable sources of energy are ones that are sustainable and will last for ever, unlike fossil fuels that are non-renewable and some will be used up in less than 100 years. That is one reason why the UK government is keen for them to be used. Another advantage is that renewables are clean sources of energy. For this reason they are often called green sources of energy because they do not release carbon dioxide and make the greenhouse effect worse. The main reason why they don't pollute is because renewables use natural resources such as weather and water. The wind turbines near my home make a whooshing noise, which annoys some people, but this is better than having the air full of sulphur dioxide fumes and acid rain like people do living next to a coal-fired power station. The government needs to meet targets from Kyoto for reducing carbon dioxide emissions. The government's view is that the more electricity that comes from pollution-free renewables, the better it is. Some people still complain about wind farms, but I like them. Most other renewable sources like wave and tidal power are still too expensive to use.

 Defining renewable, a key word in the question, is a good way to begin. An excellent, interesting explanation is given of the first part of the statement in the question (about the UK government being keen to increase the amount of energy from renewable sources). This candidate is not frightened to express his/her own opinions.

However, there is one weakness — how much is written on the second part of the statement (about only a small percentage of energy coming from renewable sources)? Only in the final sentence, when it is almost too late, does the candidate definitely begin to answer this part of the question.

The answer has many of the qualities associated with a full-mark answer, with the exception of balance between the two parts of the question statement. As a result, the marks would be limited to 5.

Summary

What mark would the answer be given?
12 out of 15. A grade-A answer.

How could the answer have been improved?
By paying more attention to the wording of questions and by double-checking answers against the questions. Most answers are very good, but if the candidate had checked back to each question before moving onto the next one, *all* the answers could have been very good.

Answer to Question 14: candidate B

(a) (i) From 68% to 29%.

 Percentages of coal in 1990 and 2000 are stated accurately from the two graphs, but the percentage reduction of 39% was not calculated. 0 marks.

(ii) From 4% to 40%, increased by 36%.

 The 36% percentage increase is not a valid alternative answer to a 10 times increase. 0 marks.

(iii) 1. Natural gas 40%, 2. Coal 29%, 3. Nuclear 25%, 4. Renewables 4%, 5. Oil 2%

 The list order is accurate and therefore would be worth 1 mark. No extra marks would be earned for stating the percentages because the question does not ask for them.

(b) Coal mines have closed and little coal remains, so that now there are only about 12 deep mines still open in the UK. More of the coal we use is now being imported, because it is cheaper whereas there is plenty of gas in the North Sea, which is brought ashore by pipeline to East Anglia. One of the main reasons for closing mines is the high cost of mining coal compared with gas drilling. Miners have to be paid high wages for working in terrible conditions underground whereas only a few men are needed on a North Sea gas rig. Coal is a fossil fuel and pollutes the atmosphere when it is burnt whereas natural gas is a clean source of energy and friendly to the environment because it is renewable.

e The style in which this answer is written is fine. Writing first about coal and then about gas, and then joining the statements together by using the link word 'whereas', is a good approach. Some of the content is less impressive, however. Not all the information relates to electricity generation, and several factual mistakes are made. For example, the UK has more coal than gas deposits. Natural gas is *not* a renewable resource; it is a fossil fuel, like coal, which means that it also emits greenhouse gases into the atmosphere, albeit in smaller quantities than coal. 2 marks.

(c) (i) HEP in the mountainous parts of the UK, such as the Scottish Highlands, where high rainfall and steep slopes are found.

e 1 definite mark for naming HEP, probably a second mark for location. Naming a place where an HEP station is located in the Scottish Highlands, such as Pitlochry or Fort William, would have made for a more precise and secure answer. 2 marks.

(ii) Renewable energy is clean and does not pollute the atmosphere, which is why governments like it. The problem is that it can only be produced where the right conditions occur. Steep slopes, high rainfall and waterfalls are best for siting HEP stations. There aren't many other places in the British Isles where new stations could be set up. That is why UK governments keep on supporting the change over to gas, which increased by 36% from 1990 to 2000, because there is plenty of it in the North Sea and it does not pollute.

e The relevant part of this answer ends with the third sentence. The part on HEP is valid and the candidate refers to both parts of the question statement. After that the candidate goes back to writing about natural gas, repeating the earlier mistake that it is a renewable energy source. The candidate could have referred to other renewables, which are increasing in importance, notably wind power for which the UK has more potential. 2 marks.

Summary

What mark would the answer be given?
7 out of 15. This would be a top grade D, close to but not quite good enough for grade C.

How could the answer have been improved?
By giving more direct answers to short questions and by showing more geographical knowledge in answers to longer questions.

 Exam tips

1 In short questions, great emphasis must be placed on accuracy. When taking answers from graphs, a slight inaccuracy, or even failure to state a unit of measurement for the correct answer, may be penalised.

2 In long questions, try to achieve a balance when two or more themes are included in the question.

National Parks

(a) What is a National Park and why were they created in the UK?   (3 marks)

(b) Study the map of National Parks in the UK.

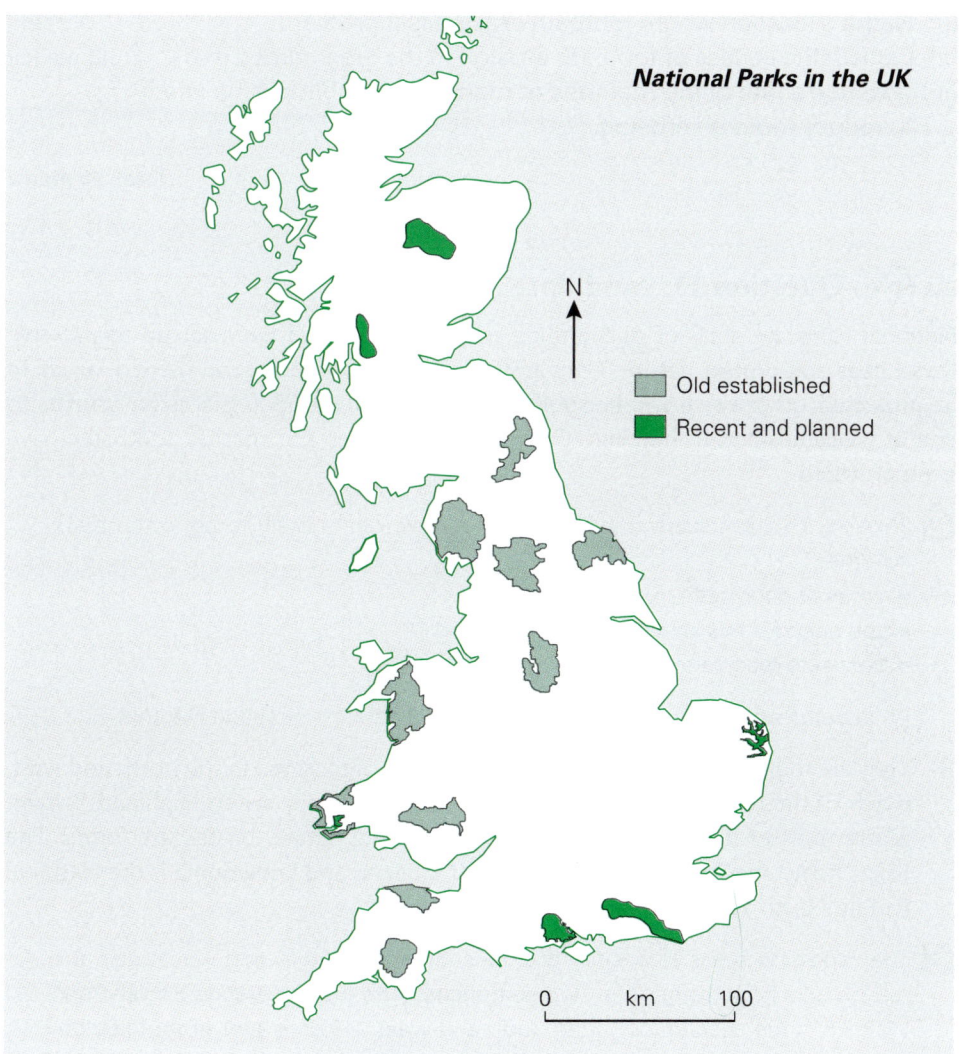

National Parks in the UK

N

Old established

Recent and planned

0   km   100

(i)   Describe the distribution of the old established National Parks.   (2 marks)
(ii) State **two** differences in location between the old established and
      the recent/planned National Parks.   (2 marks)

(c) Three land uses in National Parks in the UK are listed below:
- quarries
- forests
- reservoirs

Choose **one** of these land uses. Describe the benefits and problems of its presence in a National Park. (4 marks)

(d) Footpath erosion is a widespread problem in National Parks.
   (i)   Name a location where footpath erosion is a problem. (1 mark)
   (ii) Explain the causes of footpath erosion in the area named in (i). (3 marks)
   (iii) Describe some of the methods of management that can be used to reduce footpath erosion. (5 marks)

Total: 20 marks

Answer to Question 15: candidate A

(a) National Parks are areas of outstanding natural beauty and were set up to preserve these areas and protect nature. They are found in areas where good scenery needs to be protected, otherwise it will be spoilt. They were also set up as places for tourists to visit at weekends and in holidays. Car parks, picnic sites and marked footpaths have been provided.

e 3 marks. This candidate refers to the three elements likely to be in the mark scheme:
- areas of good scenery
- the need for this scenery to be preserved
- places for people to visit in their leisure time

An in-depth answer is not expected in a 3-mark starter question like this.

(b) (i) They are found only in England and Wales. They are located in the north and west, north of the line from the Tees to Exe. This means they are in Highland Britain, where most of the upland areas with outstanding natural beauty are found. The Lake District National Park is in northwest England and Snowdonia is the National Park in North Wales.

e The candidate starts to answer the question straightaway and would gain the full 2 marks from his/her first two sentences. (The first for stating England and Wales and the second for noting north and west, which is part of the distribution that can be seen on the map.) Even in short questions, it is good policy to elaborate further if you can, just to make certain of all the marks. Don't worry if you haven't been taught about Highland and Lowland Britain; this is not compulsory knowledge.

(ii) Two of the new National Parks are in Scotland, which had none of the old parks. This is the first difference. In England the three new parks are in the southeast of England in Lowland Britain where none of the old parks were found. This is the second difference.

e This answer shows the candidate's good level of knowledge and understanding. It is also very clear and answers the question directly. 2 marks.

(c) Quarries. These are good for people living in a National Park, because they provide jobs for local people in areas where chances of employment are often poor. Most people are farmers and there are not many other kinds of work in upland areas, where it is difficult to make a living.

Quarries are bad for people visiting the National Park because a large unsightly hole in the ground is made. This means there is visual pollution. There is also noise and air pollution from the machinery and blasting the rock with dynamite.

One example of a quarry with these benefits and problems is the limestone quarry near Castleton in the Peak District National Park.

e There is more than enough content for all 4 marks. What is so good about this answer is its organisation:
- The choice of land use is stated clearly before the main answer begins.
- Problems are described first, then benefits, and they are kept separate.
- Groups of people affected by them are named.
- A named example is stated. Even if a question does not ask for an example, referring to one always improves the quality of an answer.

(d) (i) Near Malham Cove in the Yorkshire Dales National Park.

e Another precise and accurate answer. 1 mark.

(ii) The main cause is lots of visitors. Malham is a honeypot site because nearby Malham Cove is a spectacular natural feature made out of limestone, about 100 metres high. Large numbers of people following the same track to the cove have eroded the rock away. As a footpath gets deeper, more people start walking on the grass at the side. A larger area is eroded and the footpath gets wider as well. The footpath to Malham Cove is so wide now that it looks like a road.

e The underlying cause is given and explained — visitor pressure. This is followed by explanation of how the footpath erosion occurs. The whole answer definitely refers to the location named in part (i). 3 marks.

(iii) Management means taking charge and organising work to repair footpaths. Park authorities are responsible for repairing footpaths and for making new paths. Sometimes they fence off the sides to stop people wandering off the path and making the path wider. In some places they make people follow a new route where less damage is caused. One of the main methods used is to lay new surfaces on paths in worst affected areas. This is done by making steps up steep hillsides and using wooden boards across flat swampy areas.

Of course, one way to stop footpath erosion is to stop as many people from visiting the honeypot in the first place. Park officials stop too many people going into the Peak District National Park on busy weekends by blocking the roads and forcing people to use public transport and park and ride. It is not possible to do this everywhere and many people do not like giving up their cars.

 The first statement tells us that the candidate understands what is meant by management. The rest of the first paragraph contains precise information about several methods for repairing footpaths and preventing further erosion. The second paragraph deals with a different method for managing the problem. As in earlier answers, the candidate makes specific reference to a named National Park. This is another very good answer worth the full 5 marks.

Summary

What mark would the answer be given?
20 out of 20. A* quality, with some marks to spare. These spare marks may be needed elsewhere, because even very good candidates do not find it easy to maintain this quality of answering across all the questions under examination conditions.

How could the answer have been improved?
Clearly it couldn't, but be aware that this is only one example of an answer worth 20 marks. Many equally good alternative answers are possible, for instance from candidates choosing a different land use in part (c) and a different example in part (d). Try to show some of the same strengths in your own examination answers, notably by:
- giving direct answers to the questions set
- naming places and quoting precise information about them at every opportunity

Answer to Question 15: candidate B

(a) National Parks are areas which are undisturbed by industries. They are areas with good scenery. They are for people to escape into the countryside, although some have more recently been created for the extraction of minerals.

The candidate would gain 1 mark for mentioning good scenery, but little effort is put into explaining why National Parks were created in the UK. The candidate does no more than hint at tourist visitors by mentioning people escaping into the countryside. After this, any hope of a second mark is ruined by the incorrect statement about National Parks being created for the extraction of minerals. It shows that the candidate does not fully understand. 1 mark.

(b) (i) There are three old National Parks in Wales and seven in England. None of them are near to London and none are in southeast England.

e For the first part of the answer, the candidate has relied upon counting, which does not describe the distribution. The candidate may have noticed that old established National Parks are found only in England and Wales, but does not make this clear. The second sentence would be worth 1 mark. It describes a feature of the distribution and uses an alternative way of saying that most are in the north and west. When asked to describe a distribution, you can be given credit for describing empty areas as well as those that are full. This would be another 1-mark answer.

(ii) 1. Three of the recent/planned National Parks are near London.
2. The other two parks are in Scotland.

e The candidate starts to answer the question and, by using numbered points, makes a clear attempt to find two differences. The problem is that the candidate does not establish each of the differences fully by making them two sided. (An example of a two-sided difference would be 'There were no old established parks in Scotland, but now there are two'.) Each of these answers can be described as half-mark answers, which would make a total of 1 mark for the question.

(c) Quarrying.
Benefit: it provides new jobs for the area and also provides wealth to the National Park.
Problem: the animals' habitat is destroyed and more pollution is made. The machines are noisy which will upset animals and people living nearby. Narrow country roads will need to be widened for the trucks taking away the rocks, which will be too busy with traffic.
Planting forests.
Benefit: this makes the area look more natural and prevents soil erosion. The trees will produce oxygen which will make the air in the area less polluted. The trees provide homes for animals.
Problem: whilst the trees are growing the area does not look very good. The land can be put to better uses such as farming and picnic sites.

e Instead of obeying the question and choosing **one** land use only, the candidate has given answers for two. Probably only the first answer would be marked, in which case the second choice would be crossed out and ignored, even if it were worth more. Under quarrying, there would be 1 mark for benefits ('new jobs'), and 2 marks could be given to problems (mainly for noise from the quarry and for trucks on the country roads). Some of the statements are too vague to be credited, such as 'provides wealth to the National Park' (what is meant by this?) and 'more pollution' (which is rarely rewarded unless the type of pollution is named). 3 marks.

In this example, the answer about planting forests is weaker. If it had been the only answer, it might have been worth up to 2 marks, mainly for 'prevents soil erosion' and 'provides homes for animals'. Many people would argue that the natural look of upland areas is spoilt by planting coniferous trees.

(d) (i) In honeypots that get large numbers of visitors.

 This is a general answer; no location is named. 0 marks.

(ii) When large numbers of visitors keep trampling on the land, the grass cover is worn away and bare ground outcrops on the surface. On steep hillsides, paths become so deep that after heavy rain, water flows down them and washes soil and stones away.

 Sound information is given about the causes of footpath erosion, but there is still no reference to a named location. Only 2 marks would be given.

(iii) The methods used to repair footpaths are:
- laying stones on flat surfaces
- making wooden steps up steep hillsides
- putting rafts of wood over marshy areas

 Using a list is not normally the best way to answer a 5-mark question containing the command word 'describe'. However, the candidate has made it more than just a list of methods by adding a small amount of detail about where each one is used. This part of question (d) is no longer linked to a named location, so a candidate could gain full marks for making general points. The main weakness of this answer is that only three ways are named without much elaboration, leaving at least 2 of the marks unclaimed. 3 marks.

 Summary

What mark would the answer be given?
11 out of 20. This would be a borderline grade C/D.

How could the answer have been improved?
Time was wasted in writing an extra answer that was not relevant in part (c). Part (c) is the candidate's longest answer, yet part (d)(iii) is the question with most marks attached to it. No names of National Parks or places located in National Parks are mentioned anywhere in the answer.

Exam tips

1 Refer to named places as often as possible.

2 Don't answer additional parts of questions (or additional questions) when a choice is offered. No extra marks will be gained and time will be wasted that could be used more productively, answering later questions or checking answers.

Measures of development

(a) One measure of development is GNP. What do the letters GNP stand for? (1 mark)

(b) Give a definition for each of the following measures of development:
  (i)   literacy rate (2 marks)
  (ii) infant mortality (2 marks)

(c) Study Figure 1 which shows world development. The index used is based upon high life expectancy, high rates of literacy and low rates of infant mortality. The greater the index, the higher the level of development.

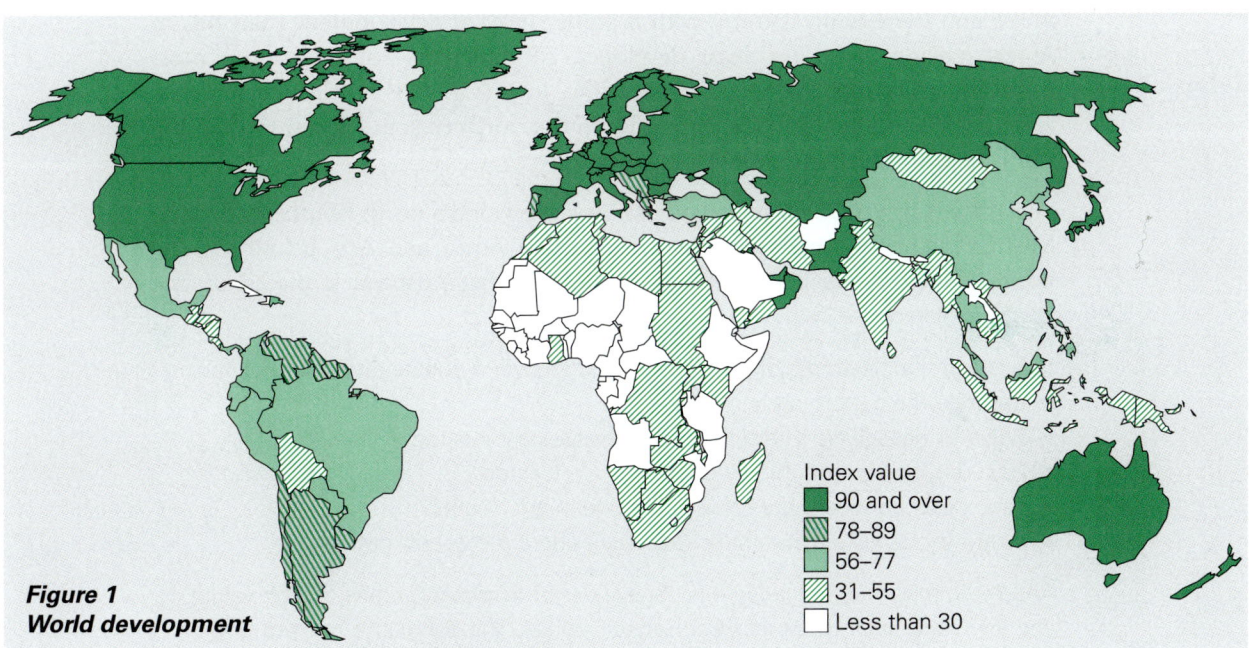

Figure 1
World development

Index value
■ 90 and over
▨ 78–89
▤ 56–77
▨ 31–55
☐ Less than 30

  (i)   Describe the pattern of world development shown in Figure 1. (5 marks)
  (ii) Name **one** country with a low level of development (55 or below) in Figure 1. (1 mark)
  (iii) Give reasons for the low level of development in the country named in (ii). (4 marks)

Total: 15 marks

Answer to Question 16: candidate A

(a) Gross National Product.

e The correct answer. 1 mark.

(b) (i) The number of people in a country who can read and write.

e The precise answer is the *percentage* of people who can read and write. The candidate showed some understanding and would gain 1 mark, but would not be given the second mark because *number* of people is not correct.

(ii) The number of children who die before they reach 1 year old per 1,000 people in a country.

e 2 marks. This is a precise and accurate answer. The unit of measurement, per 1,000 people, is correct in this definition.

(c) (i) The highest levels of development (90 and over) are mainly found in North America, Europe and Australia. The lowest levels of development (less than 30) are mainly in Africa. A lot of the countries that are quite low (31–55) are also in Africa and there is no country with a higher level of development than this in Africa, making Africa the least developed continent. In South America most countries are better developed than in Africa. In Asia there is no clear pattern, but levels of development tend to be higher in the north (e.g. in Russia) than in the south (e.g. in India).

The pattern of values on the map confirms the world's north–south divide with MEDCs mainly above 90 and many LEDCs in Africa and Asia below 55. The north–south dividing line bends southwards so that Australia is placed on the northern side of the line.

e This is a perfect example of how you should answer a question which asks you to describe the pattern shown on a map.
- Begin by describing where the highest values are located.
- Next describe where the lowest values are located.
- Then describe where the values in between are located.
- Finally try to make a summary statement about the overall pattern.

Answer in this order, because it is usually easier to pick out areas of high value and low value than those of intermediate values. The candidate noticed great variations within Asia in Figure 1, but found them more difficult to describe and locate than areas with the highest and lowest values. The low level of development in Africa is very clearly stated by the candidate. In the second paragraph the candidate's general comment is excellent and shows real understanding of the course followed by the dividing line between the rich north and poor south on a world map. 5 marks.

(ii) Ethiopia.

 1 mark.

(iii) Since the 1980s there have been many droughts in Ethiopia. It is part of the area called the Sahel and lies on the edge of the Sahara Desert. In most years the rains have failed and people have suffered from lack of food, causing famine. Thousands of people have died, especially the young and the old, who suffer most when there is not enough food.

 This is the type of answer needed, but the amount of detail given would be sufficient for only 2 of the 4 marks. The drought leading to famine that the candidate mentions is an important reason for Ethiopia's low level of economic development. However, many other reasons help to account for Ethiopia's poverty, such as high population growth, over-cultivation and firewood collection leading to desertification, as well as wars. The question asks for *reasons* and references to at least two are needed to earn full marks.

Summary

What mark would the answer be given?
12 out of 15. Receiving 80% of the marks will always be good enough for a grade A.

How could the answer have been improved?
By total precision when asked to give a definition and by reference to at least two reasons when asked to give reasons.

Answer to Question 16: candidate B

(a) Gross domestic product.

 Gross domestic product is also used in the study of development, but it is not the same as gross national product. 0 marks.

(b) (i) The % of people in a country that are literate. A person who is literate can read and write.

 2 marks. Use of the % symbol is acceptable and a full definition is given.

(ii) The % of children who die young.

 This answer lacks the precision of a good definition and the candidate would be fortunate to receive 1 mark. Infant mortality is expressed as the number per 1,000, not as a percentage. 'Young' children is not as precise as stating children under 1 year old.

(c) (i) In North America development is high. It is lower in South America with most countries between 56 and 89. It is 90 and over in Europe, which is high, but is much lower in Africa and Asia, where there are some countries with less than 30. There is no pattern to world development. For example, Australia is higher than all the other countries in the bottom half of the map.

e As the answer progresses, it becomes increasingly clear that the candidate is unable to identify the general pattern of world development from the map, high in MEDCs in North America, Europe, Japan plus Australasia, and low in LEDCs in South America, Asia and especially Africa. All this candidate does is state values for one continent at a time, continent by continent, without any recognition of the overall pattern. This style of answer is unlikely to gain more than 2 marks.

Never refer to the 'bottom half of a map' as this candidate does (nor to top, left or right of a map). This suggests a terrible lack of geographical understanding; always use compass directions instead.

(ii) Africa.

e Africa is a continent, not a country. This mistake is made by many examination candidates. Make sure you are not among them. 0 marks.

(iii) In Africa there are many droughts, e.g. in the Sahel. The Sahara desert is growing bigger. This has been caused by the many droughts and desertification. Desertification is when the soil is blown away from the surface, leaving only sand and rock. Africa is a place that has many wars. There is always fighting, leading to refugees and people starving in refugee camps. This is why the country of Africa has a lot lower level of development than the UK.

e The quality of this answer is badly affected by the fact that the candidate considers Africa to be a country. Although the candidate names the Sahel region in Africa, no African country is named in the answer. Therefore, an examiner could only treat this as a 'general' answer, not specific to a named country. Some marks could still be given for general points that are relevant to the question theme, but it is unlikely that the candidate would be given more than half marks. If a country like Ethiopia had been named by the candidate, this would have been a much better answer than the one given by candidate A because of references to drought, desertification and wars. Without a named country it is worth the same. 2 marks.

Summary

What mark would the answer be given?
7 out of 15. Grade D.

How could the answer have been improved?
By using a better technique to describe the pattern from the world map, and by knowing the difference between continents and countries.

Exam tips

1 Know the locations of the six inhabited continents. This is essential knowledge for interpreting world maps and understanding world themes like economic development.

2 Being precise and accurate is vital when giving definitions.

3 When candidates are required to base an answer upon a named country, some marks will be given to relevant general answers. However, a good choice of a named country is essential if you want to earn high marks.

Forest destruction

(a) Study the photograph below which shows an area where tropical rainforest has been cleared.

   (i)   What is the main land use?   (1 mark)
   (ii) State evidence from the photograph that suggests that this area used to be forested.   (2 marks)
   (iii) Is the land use shown in the photograph sustainable? Explain your views about this.   (5 marks)

(b) When rainforests are going to be cleared, there are many conflicts of interest between different groups of people. The diagram below shows some of the groups of people with interests in natural rainforests.

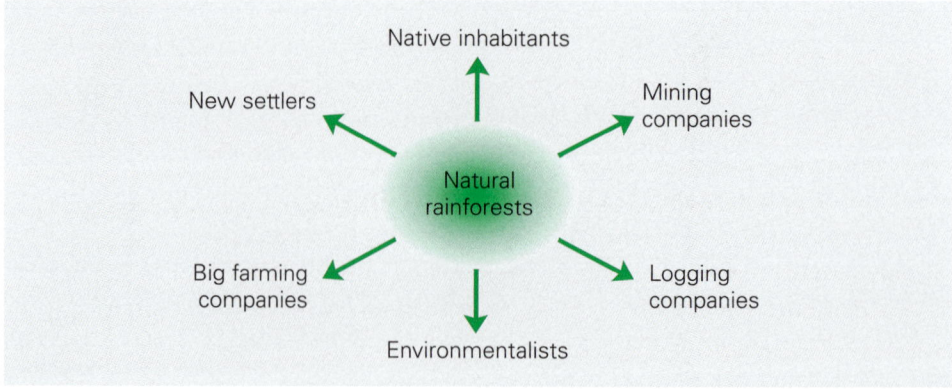

(i) Name one group of people likely to be in favour of forest clearance and one group of people likely to be against forest clearance. **(1 mark)**

(ii) Explain why there are conflicts of interest between the two groups named in part (i). **(6 marks)**

Total: 15 marks

Answer to Question 17: candidate A

(a) (i) A field with cattle in it. The cattle look like beef cattle.

e The photograph is used and the main land use is stated.1 mark.

(ii) Trees can still be seen on the left side of the photo. They do not look like trees that have been replanted. The large tree in the middle also looks like one that was left when the land was cleared.

e The candidate's answer is again based on observation. It refers to points on the photograph where evidence can be seen and makes sensible suggestions in line with the question asked. This answer demonstrates the correct approach to answering photograph-based questions. 2 marks.

(iii) When rainforests are cleared, there are fewer trees for coverage of the soil so that when it rains there is no interception. The rain hits the soil, causing soil erosion. Nutrients in the soil are washed away and leached too so that the soil loses its fertility.

This will happen in the area shown on the photograph because there is now a large area without trees. Cattle will trample over the land and make soil erosion more likely. Therefore, I do not think that the photograph shows a sustainable type of land use because the land will lose its fertility. Sustainable means being able to keep using the land for ever without losing its fertility.

e In the first paragraph the candidate gives a general answer about the dangers which arise from clearing rainforests. If no further details were added, this would have been an example of the first level of answer and worth 2 marks.

The second paragraph greatly improves the quality of the answer, because the candidate refers to the photograph again and shows a clear understanding of what sustainable means. This would take the answer into the next level of answering and make it worth 4 marks.

Why not all 5 marks? The candidate fails to think about the opposite viewpoint — that the farming in the area shown on the photograph might be more sustainable than some of the other land uses in places where rainforest has been cleared. For example, there is a good covering of grass which is protecting the soil, and considering the size of the field, the number of cattle in it is unlikely to cause problems of overgrazing.

(b) (i) In favour of forest clearance — logging companies.
Against forest clearance — environmentalists.

e Two likely groups are selected. 1 mark.

(ii) The reason why logging companies are in favour of forest clearance is that they know that they can make a lot of money by cutting down the big forest trees. In the Amazon it only costs $1,000 to cut down a mahogany tree but the loggers sell it for $10,000. The demand for hardwood trees for furniture is great in MEDCs. It is easy to do and cheap for the logging companies, because the forest is a natural resource. Each big tree cut down gives a large amount of wood. Other examples of trees logged from tropical rainforests are teak and ironwood. Logging companies care more about making profits than about the environment. A lot of logging has taken place in Indonesia and Brazil.

The reason why environmentalists are against forest clearance is that it causes global warming. By burning trees that are not needed for logs, carbon dioxide is released into the atmosphere. This traps in the heat from the surface, which is called the greenhouse effect. When the trees are destroyed this also means that there are fewer trees left in the world to absorb the carbon dioxide. The world's supply of oxygen may be reduced as well by removing some of the world's largest remaining areas of forest.

e In the first paragraph the reasons why logging companies are in favour of forest clearance are stated in a clear and precise manner. In the second paragraph, environmental concerns about increasing global warming are covered in a reasonably effective way. But the candidate does *not* address the central theme of the question — the conflicts of interest between the two groups. The candidate treats the attitudes of the two groups separately; the closest the candidate comes to linking the two is in the comment towards the end of the first paragraph that 'logging companies are more interested in making profits than about the environment'. As a result of placing insufficient emphasis on conflicts, the answer would only gain 5 of the possible 6 marks.

Summary

What mark would the answer be given?
13 out of 15. A comfortable grade-A performance. The basic content in the answer is very good and the photograph is well used in part (a).

How could the answer have been improved?
By adding extra comment, to convert what is already a good answer into the best possible answer. In (a)(iii) some comment supporting sustainable use would have helped, even though it was quite reasonable to take the view that the land use shown is not sustainable. When asked for views, as in (a)(iii), try to give views from both sides, even if you think one is much better than the other. When asked for conflicts

of interest, as in (b)(ii), make very clear why the interests of the two groups of people conflict by linking together comments for the two groups.

Answer to Question 17: candidate B

(a) (i) Ranching.

e 1 mark. The answer '*cattle* ranching' would have made it more obvious to the examiner that observation from the photograph was being used.

(ii) Some trees can still be seen on the photograph.

e 1 mark. One short statement is never going to be enough for 2 marks. The candidate needs to study the photograph in more detail and use more evidence from it.

(iii) I think that the farming is not sustainable. The natural environment is destroyed as people clear the land for farming due to the rains of the tropical rainforest. The soil becomes useless for farming as nutrients are leached out by heavy downpours. The ground becomes unfit for farming and the farmer needs to move on to a new area while the old area becomes a desert and nothing will grow. This process is repeated again and again.

e This is a general answer about what happens after rainforests are cleared; there is nothing anywhere in the answer to indicate that the candidate is being influenced by what can be seen on the photograph. Indeed, comments that 'the farmer needs to move on to a new area' suggest that the candidate is referring to slash and burn subsistence farming, instead of to the commercial cattle rearing shown on the photograph. It would just reach the first level in the mark scheme. 2 marks.

(b) (i) Big farming companies and native inhabitants.

e The candidate does not make it clear which group is which. The examiner might be willing to assume that they are being stated in question order, that big farming companies is the group in favour and native inhabitants is the group against. If so, the answer would be worth 1 mark. However, it is risky to give a brief answer that is unclear.

(ii) The more land that is cleared, the larger the area of land that can be used for farming. Big American burger companies set up cattle ranches so that they can supply their restaurants with beef. It is cheaper to rear cattle in LEDCs in the tropics where there are large areas of land for grazing and governments are keen to make money out of exporting, whereas the native people lose their homes in the forests. They live from hunting and collecting forest products such as rubber and drugs for medicines. Without the forest, they cannot survive and this is why clashes arise between these two groups. They are often forced off their land.

One important way in which this is a better answer than the one given by candidate A to the same question is that this candidate clearly establishes why conflicts of interest exist between the two groups. Although the big farming companies are dealt with first, the candidate uses the important link word 'whereas' before referring to the native inhabitants, and then adds comment on the reasons for clashes between the two groups. However, candidate A's answer was fuller and more accurate; the comment in candidate B's answer about the big foreign burger companies setting up ranches is not generally true. Overall, this answer would be worth 5 marks because of the good focus on answering the question set.

 ## Summary

What mark would the answer be given?
10 out of 15. Grade C.

How could the answer have been improved?
By making greater use of the photograph and what it shows in all parts of (a).

Exam tips

1 Always make a detailed study and full use of a photograph. Make specific references in your answer to what can be seen and where it can be seen.

2 When asked for your views in a question, or for explanation about conflicts of interest, add comments in your answer to show the examiner that you are giving a direct answer to the question set.

3 Before taking the examination, make sure that you understand important words like sustainable. For example, candidate A's inclusion of a definition of sustainable in the answer to (a)(iii) helped.

Acid rain

(a) Study the diagram below which shows one cause and one effect of acid rain pollution.

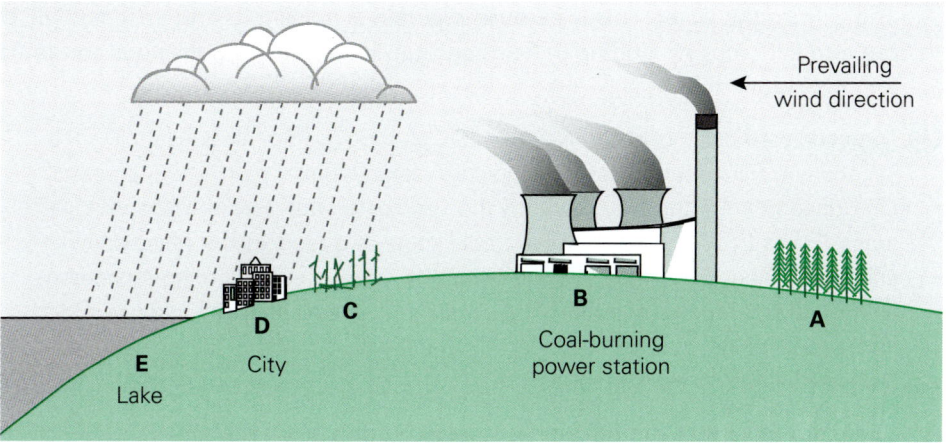

(i)   State the cause and the effect of acid rain shown on the diagram.   (2 marks)

(ii) Why is the effect likely to be greater at C, D and E than at A?   (2 marks)

(iii) Describe the likely effects of acid rain in the areas labelled D and E.   (5 marks)

(b) (i)   Name **one** strategy for reducing the amount of acid rain produced.   (1 mark)

(ii) Explain how it works.   (3 marks)

(iii) How successful has it been? Explain your answer.   (2 marks)

Total: 15 marks

Answer to Question 18: candidate A

(a) (i)   The cause is the coal-burning power station with the pollution from it carried by the prevailing winds. The effect is the dead trees at C.

e 2 marks.

(ii)   The pine trees look healthy at A and dead at C, because the fumes from the power station chimneys are being carried towards C by the prevailing wind. Acid rain is caused by burning fossil fuels, which release oxides of sulphur and nitrogen into the air. These are coming out of the chimney at the power station. The prevailing wind is the one which blows for most of the year. When it rains, the acid from the fumes is carried down to the ground. Therefore there will be acid rain at C, D and E, but less acid rain at A.

e A clear and accurate answer. The candidate shows that he/she understands both prevailing wind and acid rain, and gives much more information about the formation of acid rain than can be expected in a 2-mark answer. 2 marks.

(iii) Sulphur dioxide falls from clouds as precipitation in the form of acid rain. Acid rain can corrode limestone and cement, and so it can destroy buildings and statues (area D). It can also destroy wildlife by getting into rivers and lakes (area E). This can kill the fish and other creatures living in lakes, as well as killing anything that drinks from the water. This is because acid rain is poisonous and so it can kill. It can also kill crops by landing in fields and can leave the soil poisonous and unfit for cultivation.

e The candidate describes the likely effects, as required by the question. In the second and third sentences the candidate makes several mark-earning points. One good point is the way in which the candidate separates the answers for areas D and E. After this, however, little more is added to the basic answer and 'killing' and 'poisonous' are overused. The references to farming at the end are not relevant to this question. The candidate loses his/her way and gives the impression of not having sufficient knowledge to sustain a full answer to a 5-mark question. 3 marks.

(b) (i) The main thing that can be done is to stop burning fossil fuels.

e This is a valid strategy. 1 mark.

(ii) Nuclear power could be used instead. It does not give out sulphur dioxide and so does not produce acid rain. This is a good example of an alternative source of energy, and this is what must be found — alternative sources of energy. Other sources could be the likes of solar power as well as hydro and wind power. They have the advantage of being renewable and green sources of energy. This means that all of them are pollution free; there is no air pollution as they use natural features such as sun, water and wind.

e These references to energy sources that are alternatives to fossil fuels match the strategy named in part (i). Four alternative types of energy are named and there is sufficient detail for all 3 marks to be earned.

(iii) These sources are being used but in very small quantities when compared to that of fossil fuels. These alternative sources do not get used much because they are expensive to set up and it is difficult always to get enough of an energy yield (as in the case of solar power in the UK). Nuclear power also has many dangers and health risks because it is radioactive. Wind power and water power can create visual pollution too.

e In the first sentence the candidate implies that alternatives have not been very successful as replacements for fossil fuels. To support this view, expense, low energy yield, risks and visual pollution are mentioned. These add up to more than enough for full marks in a 2-mark question, but it is better to give too much rather than too little in answers, provided that you are not running short of time. 2 marks.

Summary

What mark would the answer be given?
13 out of 15. A comfortable grade-A answer.

How could the answer have been improved?
By providing more details of the effects of acid rain in (a)(iii), for which more knowledge was probably needed.

Answer to Question 18: candidate B

(a) (i) Dead trees at C are the cause of acid rain pollution. The effect comes from the prevailing wind direction that is shown on the diagram.

 Dead trees at C is one effect of acid rain pollution, not the cause of acid rain. The real cause is coal being burnt in the power station and releasing oxides of sulphur. The candidate has misinterpreted cause and effect in relation to acid rain. 0 marks.

(ii) The prevailing wind, clouds and rain are all affecting C, D and E. The forest at A is on the sheltered side and gets less wind.

 For most of the answer the candidate is stating rather than explaining, and not always accurately; the forest at A, for example, is not on the sheltered side. Although the candidate seems to imply that the prevailing wind is responsible, he/she does not succeed in explaining why the effects are likely to be greater at C, D and E. 0 marks.

(iii) Acid rain affects buildings because it is corrosive. The brickwork on the outside is destroyed. Buildings built out of limestone are affected most because limestone rock dissolves in acid rainwater. Statues are also affected and after many years they crumble and you cannot see who they are. It costs a lot of money to clean and repair buildings in big cities like London, e.g. the Houses of Parliament, which is made of limestone.

Lakes are part of the hydrological cycle. Acid rain is part of a vicious cycle. As lakes get polluted, lake water is evaporated and falls again as acid rain, which kills trees, plants, animals and fish. Lakes can become so acidic that all the water creatures die.

 To the candidate's credit, every effort has been made to give a full answer, which is important in a 5-mark question. In the first paragraph the candidate makes some good points about the effects of acid rain in a city; references to

limestone and mention of an example are useful. The second paragraph is not as good. It confirms what became clear in the previous answer to part (ii) — that the candidate does not really understand how acid rain is formed. One effect is still included, namely killing fish and other water creatures. 4 marks.

(b) (i) One way of helping to prevent as much acid rain is by using filter systems placed in the chimneys.

e Using filters is a valid strategy. 1 mark.

(ii) They are set up in chimneys to control the amount of pollution that goes out. If less pollution is let out, there will be less pollution in the hydrological cycle and less acid rain.

e This answer does not go much further than the one given to part (i). The reference to setting up the filter in the chimney could be given 1 mark. The candidate needed to give more details, for instance about 'scrubbers' (FGD equipment), which can remove up to 95% of sulphur dioxide emissions using limestone. Would the choice of another strategy, such as alternative sources of energy used by candidate A, have given more opportunities for a longer answer with more content? 1 mark.

(iii) Not very successful. This is a costly procedure, which is why it is only used in a few places. Unless power companies are forced into doing it by the government, they do not bother. It would not stop all the pollution getting into the atmosphere either. So why go to the expense? It might be better to look for alternative fuel sources instead, but these are more expensive as well.

e The candidate gives two reasons why the strategy has not been successful, namely cost and not stopping all the pollution. The cost problems are detailed best. The answer is well focused on the question set, because the candidate continues to use the same strategy named in part (i). 2 marks.

Summary

What mark would the answer be given?
8 out of 15. This would be a borderline grade-C answer.

How could the answer have been improved?
By greater consistency in answering the different parts. One of the features of many grade-C answers is a mixture of good answers, like (a)(iii), and weak answers, like (a)(i),(ii) and (b)(ii). To be certain of securing a grade C or higher, good answers need to dominate more than they do in this answer.

Exam tips

1 Try not to lose marks on the short, source-based questions, as in (a)(i) and (ii). These are often the easiest questions.

2 Don't make mistakes when asked to distinguish between causes and effects.

3 When asked to choose one item or one place, or in this example one strategy for reducing acid rain, always choose the one that you can write most about. By choosing alternative energy sources rather than filters in chimneys in part (b)(i), the probability is that candidate B would have had more to write about in (b)(ii).

Global warming

(a) Study the graph below which shows average world temperatures since 1860.

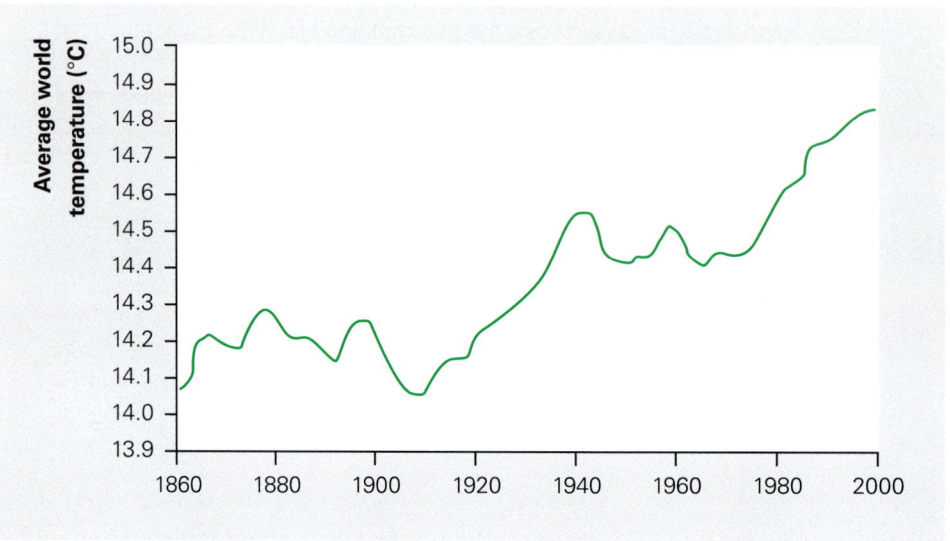

Describe how the graph suggests that global warming is happening. (3 marks)

(b) Explain the greenhouse effect in relation to global warming. (4 marks)

(c) (i) Describe **two** possible disadvantages of global warming for people. (4 marks)

(ii) People living in some countries, and in certain areas within countries, are more worried about the possible effects of global warming than others. Suggest reasons for this. (4 marks)

Total: 15 marks

Answer to Question 19: candidate A

(a) The average temperature was about 14.1°C in 1860, but by 2000 it had risen to more than 14.8°C, an increase of 0.7°C. Although there are fluctuations in the graph, the trend has been upwards. Greater increases in temperature are seen after 1920 and after about 1970. It certainly shows that the world is warming up.

e 3 marks. This is a good answer with several strengths, and it illustrates well what you should do when describing a graph:
- The general trend has been noted — upwards.

- Key values have been *stated* (14.1 and 14.8°C) and *used* (difference of 0.7°C).
- Variations within the trend have been spotted — the two times when the temperature line rises more steeply.

(b) The sun heats up the land surface. Some of this heat is trapped by the air in the atmosphere above us, while the rest is lost into space. The underlying cause of the greenhouse effect is that the amounts of greenhouse gases in the atmosphere have increased. There is a thicker layer of carbon dioxide in the atmosphere as a result of factories and cars burning off fossil fuels. This carbon dioxide traps some of the heat which was previously lost into space, making the earth warmer, like the graph shows. In addition, more heat is coming in from the sun, because the hole in the ozone layer is allowing this to happen, and this is helping to warm up the earth as well.

e Apart from the last sentence about the effects of the hole in the ozone layer, which is irrelevant and wrong, this answer demonstrates an accurate understanding of the greenhouse effect. It is also clearly stated. The last sentence would be ignored by the examiner, because it does not contradict the correct part of the answer. Even so, it would still have been better if the candidate had not written it, because it does suggest some misunderstanding and confusion. 4 marks.

(c) (i) One bad effect is the rise in sea level. This will lead to flooding in lowland areas, like the Ganges Delta in Bangladesh, where millions of people live and grow rice on fertile silt soils.

Some people are predicting that the amount of bad weather is going to increase. There will be more storms, hurricanes etc., which will kill more people and ruin their farmland. There are already a lot of cyclones in Bangladesh and these may become more frequent and stronger, which is bad news for the people of Bangladesh.

e The candidate identifies two possible disadvantages and elaborates upon them; by elaborating the candidate obeys the command word in the question, 'describe'. The reference to a named place improves the quality of the description. This answer would be comfortably worth all 4 marks.

(ii) People living in Bangladesh are very afraid about what is going to happen to them. They are already very poor, so that if global warming stops them growing as many crops, or if more cyclones destroy their crops, they are going to be badly affected. It is obvious that people living in coastal areas will be worst affected by global warming.

e This answer fails to match the high quality of the earlier ones. The candidate begins well by continuing from part (c)(i), but doesn't manage to take the answer much further. It is no use stating that 'it is obvious' without explaining how or why. The candidate would have done better to have compared the situation of people living in low-lying coastal areas with that of those in higher areas above the flood danger level, either in coastal or inland regions. 2 marks.

Summary

What mark would the answer be given?
13 out of 15. Grade A.

How could the answer have been improved?
By giving a two-sided answer in the final part; adding reasons for areas less affected would have done this.

Answer to Question 19: candidate B

(a) The graph suggests that global warming is happening because the temperature in 2000 is higher than in 1860. The lowest temperature, however, was just after 1900. The highest is in 2000.

 The candidate would earn 1 mark for suggesting that temperatures have indeed shown an increase. However, no values are used to support the answer and so no further marks would be earned. The other point made about the lowest temperature after 1900 is not significant for the question set. 1 mark.

(b) I am going to explain how the greenhouse effect works. The hole in the ozone layer lets more of the sun's heat in. The Earth is becoming hotter because of this. This is global warming. Greenhouse gases like CFCs trap the heat and this is making the Earth's surface hotter than it used to be. The ban on using CFCs in aerosols and fridges should reduce the greenhouse effect, but it will take a long time.

 This answer shows that the candidate is confused between the greenhouse effect and the effects of the hole in the ozone layer. In fact, more of the answer is about the ozone hole than the greenhouse effect. However, it is true that CFCs are one of the greenhouse gases, although they are not as important as carbon dioxide. Therefore, the sentence beginning 'Greenhouse gases...' is of some relevance. 1 mark.

(c) (i) One disadvantage of global warming for people is the increase in skin cancers and eye diseases. This is because of the great heat of the sun as less of the sun's ultra-violet rays are filtered out. It is affecting people's health badly.

A second disadvantage of global warming for people is flooding of houses and farmland as sea levels rise. People living on islands and in river estuaries will be flooded first and worst.

 The first disadvantage, the effects on health, is caused by the hole in the ozone layer and is not relevant. This shows that the candidate is still confusing the ozone hole and greenhouse effect. The second disadvantage does relate to global warming, so is relevant, and would gain 2 marks.

(ii) People living in coastal areas are more worried about the possible effects of global warming than others. In the UK, people living in the Fens are in big danger. Some people think that the Thames flood barrier will not be big enough to save London, whereas people living in the Highlands of Scotland have no worries about flooding. People in Bangladesh live on the flat lands of the Ganges delta. Many are already flooded out every year, so the problems for them can only get worse with rising sea levels. In nearby India many people live in the Ganges valley but the land is not as low as in the delta. Flooding in India is not expected to be as bad.

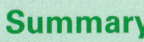 This is a much better answer because it is entirely focused on possible effects of global warming (and the ozone layer is not mentioned once). Another positive feature is that it is based upon named places, both those likely to be affected and those likely to be left unaffected by rising sea levels. 4 marks.

Summary

What mark would the answer be given?
8 out of 15. This would be a grade C.

How could the answer have been improved?
By spending more time answering the questions set:
- By concentrating upon global warming and greenhouse effect only, and ignoring the irrelevant hole in the ozone layer.
- By not wasting time writing out the question; this was done at the beginning of every answer.

Exam tips

1 Make sure that you know the differences between global warming and the ozone hole before going into the examination. Many marks are lost in examination answers because candidates confuse these two types of pollution (as both candidates do, although B more than A).

2 Don't waste time and space by repeating the question in your answer. Start answering the question straightaway.

Wilderness areas and their development

(a) Study the photograph below, taken in Antarctica, the largest area of wilderness left on Earth.

The definition of wilderness is 'an area of undeveloped land which is still natural'. Describe how the photograph shows that Antarctica is a wilderness.

(5 marks)

(b) Deposits of oil and other minerals are known to exist in Antarctica. At present international treaties prevent all countries from drilling and mining in and around Antarctica. Is this the best strategy? Write down and explain the views that different people might have about this, as well as expressing your own view.

(7 marks)

Total: 12 marks

Answer to Question 20: candidate A

(a) The photo shows an area with no houses, roads or any other signs of human settlement. This fits with the definition that the land is undeveloped. What is shown in the photo are penguins, blocks of ice and loose stones. At the back of the photo there is a glacier with lines of moraines, surrounded by bare rocks, and it looks mountainous. It must be too cold and rugged for people to live here. That is why it is still natural, i.e. no signs of people, just penguins!

e 5 marks. The candidate gives exactly the type of answer expected for a question with the command words 'describe how':
- What can be seen in the photograph is described.
- At two points in the answer the candidate comments upon how his/her observations show that the area has the characteristics of a wilderness.

(b) Preventing countries from drilling and mining in Antarctica seems to be the best strategy to me. If they were allowed to drill for oil in the area shown in the photo, it would be a disaster for the environment. An oil rig would destroy the views and the noise from drilling would disturb the penguins. Then think what would happen if there was an oil spill. The penguins would be covered in treacle–like black oil, like we have seen happen to sea birds in the UK when there has been an oil spill. There wouldn't be any rescue services to help them as there was in south Wales when a oil tanker hit rocks near the entrance to the harbour at Milford Haven a few years ago. The spill affected tourist beaches in Tenby further along the coast, because currents carry oil slicks long distances.

Environmentalists will support the strategy of preventing mining. They would say that Antarctica is one of the few real wilderness areas left in the world. Why pollute it as has been done in most of the rest of the world? I presume that oil companies would be against the strategy of no drilling. They would argue that the world needs new supplies of oil, but they would say this mainly so that they could make a good profit.

My view is that governments should spend more money looking for alternatives to oil so that global environmental problems such as the greenhouse effect are reduced. What can be seen in the photo looks too good to be ruined. Mining and drilling here would not fit the need for new developments to be sustainable.

e 7 marks. The candidate has fulfilled all the question required, by expressing views for and against the strategy and by making his/her own view clear. There are three particular strengths in this answer:
- The candidate continues to look at and use the photograph throughout the answer.
- Reference to an actual example of what happened off south Wales is included.
- The candidate recognises that future mining in Antarctica would not be sustainable development.

Summary

What mark would the answer be given?
12 out of 12. Unless other answers from the candidate were considerably inferior, this candidate would be well on the way to an A*.

How could the answer have been improved?

It couldn't in terms of what can be expected from an answer at GCSE level, even though other able candidates might explore the views of other groups of people more fully than this candidate does.

Answer to Question 20: candidate B

(a) There is nothing to show that people live here — no farms or houses, no roads or other ways of reaching the area. There is no pollution. That is why it all looks so natural with rocks and ice.

> *e* 2 marks. The main problem is that the candidate makes no attempt to describe in a positive manner what can be seen in the photograph until the last sentence. Most of the answer is about what is not there and negative statements are inferior to positive ones. The amount of content is well short of that expected in a 5-mark question.

(b) I think the best strategy is not to mine in Antarctica and that it should be left as it is. Mining causes environmental problems because it destroys beautiful scenery and then leaves big surface holes and waste heaps of rock behind when mining finishes. The mining company would need to build a settlement for workers, port and roads before mining operations got started, and these would ruin the natural appearance of a place like Antarctica. It is a long way from where anyone lives, which would make it expensive for mining companies to operate there. The mining companies could go bankrupt and just abandon operations, leaving ghost towns behind them. These would not be a pretty sight. The world has plenty of oil, so I cannot see why oil from Antarctica is needed, and I am sure that a lot of other people would agree with me. I know environmentalists would. For all these reasons I think that it is the best strategy not to mine in Antarctica.

> *e* The candidate gives his/her personal view and supports it reasonably well by referring to some of the problems caused by mining. The main problem is that the candidate does not explore the views of other people. This would put the answer into the middle category, good in parts, and it would gain 5 of the 7 marks.

Summary

What mark would the answer be given?
7 out 12. Marginal grade C.

How could the answer have been improved?
By a fuller description of the photograph and by consideration of the views of others as well as the candidate's own.

Exam tips

1 When a photograph is supplied, make maximum use of it and describe it in positive terms.

2 When different views are required, support them with evidence from case studies or resources provided with the question.